THE WOODLOT MANAGEMENT HANDBOOK

Stewart Hilts ✻ Peter Mitchell

illustrations by Ann-Ida Beck

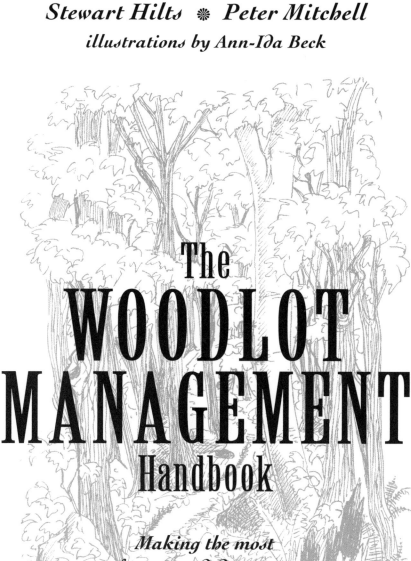

The
WOODLOT
MANAGEMENT
Handbook

*Making the most
of your wooded property
for conservation, income
or both*

FIREFLY BOOKS

A Firefly Book

Published by Firefly Books Ltd. 1999

Cataloguing-in-Publication Data

Hilts, Stewart

The woodlot management handbook

ISBN 1-55209-236-4

1. Woodlots - Management. I. Mitchell, Peter II Title.
SD387.W6H547 1999   634.9˙9   C98-932816-0

The authors have tried to be as accurate as possible. We regret any errors or omissions.

Design: Counterpunch/David Vereschagin
Editor: Meg Taylor
Copy editor: Sandra Tooze
Production: Denise Schon Books Inc.
Index: Barbara Schon
Photographs: Stewart Hilts and Peter Mitchell

Published in Canada in 1999
by Firefly Books Ltd.
3680 Victoria Park Avenue
Willowdale, Ontario
Canada M2H 3K1

Published in the United States in 1999
by Firefly Books (U.S.) Inc.
P.O. Box 1338, Ellicott Station
Buffalo, New York
USA 14205

Printed and bound in Canada by Friesens, Altona, Manitoba

Printed on acid-free paper

Second Printing, 2002

CONTENTS

INTRODUCTION

The woods holds a special place in the landscape of eastern North America. Whether you are a woodland landowner, a would-be woodland owner, or just someone who enjoys walking in the woods, you probably understand their special mystique.

All our lives we have enjoyed walking in the woods. First with our parents, then with our own children, and now often with students or other woodlot owners, we simply enjoy the beauty of being in the woods.

Walking in the woods, you can watch for trees, wildflowers, mushrooms, animals, and birds. You can listen to the frogs in the wetland, or the wind in the treetops. In the spring, the miracle of growth is all around you; in the fall, the spectacular colors blaze; and in the winter, you can follow the tracks of unseen woodland wildlife. The woods is a place of peace in our hectic modern world, where you can get away from it all, however briefly.

The woods is a magical place.

At the same time, the woodland is a place where what is good ecologically is usually also good economically. Careful management of woodlots can protect the environment while also providing a significant economic return to the landowner. In fact, the greater care given to environmental sustainability, the higher the economic return will probably be. This is a paradox many woodland owners probably haven't figured out, but we'll explain it below.

However, reaping ecological and financial benefits depends on a good understanding of the principles of woodlot management. Through this book we hope to introduce you to a better understanding of the woodlands of eastern North America, and to an understanding of woodland management based on the principles of ecology.

The principles described in this book cover woodlot management in a large geographic area of eastern North America, as shown in the map below. This region, including the eastern deciduous and the northern hardwood forests, verges into boreal forest to the north, prairies to the west, and into the southern pine ecosystem to the south.

Within this region there is, of course, enormous variation in woodlot types, in tree species, and in forest communities. This, in turn, results in different woodlot products that can be harvested, different growth rates, and different levels of financial return.

In this book we will start with a basic understanding of woodland ecology. We will help you understand the dynamic process of changes that occur over the seasons, from the dormancy of winter, through the reawakening of spring, to the growth of summer.

We will explain the details of relationships between the soil, the water, and the climate, and the relationship of these physical factors to the biological life—trees, plants, and wildlife—that make up the woodlands.

We will outline how to build a working description or inventory of your woodland as the foundation for management.

Several different options for woodland management will be discussed. These range from management choices designed to conserve the natural ecosystem to management for sustained timber harvest. We also include how to create or recreate new woodlands through reforestation or through allowing natural regrowth.

We believe that woodlot management, or "woodland steward-ship" as we are wont to call it, should reflect the importance of the particular woodlot within the broader landscape, as well as the landowner's own vision for it. We emphasize an understanding of all management options as a basis for this choice, leaving it up to you, the landowner, to make your own choice in the end.

We include a brief discussion of how to develop your own woodland-stewardship plan, and a chapter on what to look for if you are hoping to purchase your own woodland. We also provide you with some ideas for long-term conservation: how you can hand your woodland down to the next generation and still ensure that it will be cared for.

Finally, the book includes a list of sources to assist you in apply-ing these ideas directly to your own land.

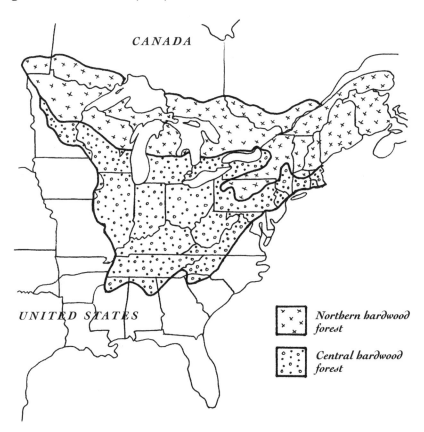

The deciduous forest regions of eastern North America.

CHAPTER 1

WHAT THIS BOOK IS ALL ABOUT

We first must have a biologically sustainable forest before we can have an economically sustainable yield (harvest) of any forest product, be it woodfiber, water, soil fertility, or wildlife … we must first practice sound "bioeconomics" (the economics of maintaining a healthy forest) before we can practice sound "industrio-economics" (the economics of maintaining a healthy forest industry), before we can practice sound "socio-economics" (the economics of maintaining a healthy society). It all begins with a solid foundation – in this case, a biologically sustainable forest.
— C. Maser, *The Forest Is a Living Trust,* 1990.

For fifteen years now, we have been developing programs to help rural landowners understand and care for their properties. We know from this work that rural landowners love their land and are always ready to learn more about it. The single most common group of questions landowners have are related to caring for their woodlands and creating new woodlands by planting trees.

Some owners want to leave their woods completely alone, to preserve it just for the birds. Others need or want to harvest timber or firewood on an ongoing basis. In between are owners interested in nature study, hiking, hunting, and other activities. Many landowners choose a combination of these relationships with their land.

As long as the care of your woodland is "sustainable" – that is, it leaves the woods in healthy ecological condition – we believe that management should be the choice of the landowner. This is why we use the term "stewardship" to refer to this role of landowners. In our view, stewardship simply means the care that private landowners give to their land. It implies some active management based on understanding and an ethical commitment to leave your land in as good condition as, or better than, you found it.

In this book, we will be reviewing a full range of woodland-management options, from preserving your woodland for nature to sustainable timber harvesting. Management that takes into account this complete range of options is sometimes referred to as "holistic forestry."

Woodland Stewardship

Over the past decade, the science of forest management has changed substantially, from a historic emphasis on timber and some wildlife management to an emphasis on caring for the entire forest as an "ecosystem."

An "ecosystem" is the sum total of all the factors and components that make up the natural system in a given region. It includes physical factors such as the soil, water, sunlight, nutrients, and energy that enable the system to function. It includes the plants that grow in the area to form a plant community. And it includes the wildlife that lives in the area, from birds to mammals to the millions of insects that we rarely notice.

Above all, the term "ecosystem" emphasizes the relationships and interactions among all these components. Thus an ecosystem is a complex web of individual parts, all interacting with each other, as we will discuss in more detail in Chapter 2, on woodland ecology.

A new term closely associated with our understanding of ecosystems is "biodiversity." This term has grown out of international concern for the disappearance of species as humanity eliminates more and more of the remaining natural habitats on the planet. It refers to the complete biological diversity in a region, including

- the diversity of plant communities in a landscape,
- the diversity of species in a community,
- the diversity of genetic characteristics in a single species, and
- the genetic traits of an individual organism.

Conservation of biodiversity is now a global priority, another issue to which we will return in the following chapter.

But woodlands have other important functions as well. Conserving water resources is one of the most critical; forest vegetation plays a key role in the hydrological cycle, moderating runoff and minimizing erosion. Woodlands also provide our most important wildlife habitats and are widely used for recreation, to say nothing of their economic value.

When we speak of ecosystem management, or conservation of biodiversity, we are placing the emphasis on the woodland as a whole, not just the trees and certainly not just those trees that we might harvest for timber. This new emphasis on the big picture – the entire woodland in all its complexity – is one half of the idea of holistic forestry, or what we refer to as woodland stewardship. It recognizes all the ecological functions that woodlands perform.

At the same time, people are usually also part of this picture. At least in the landscape of eastern North America, most woodlands are privately owned, and these landowners use their woodlands for a wide range of purposes. These purposes can be described as the values that woodlands provide to society, though you can also argue that they have their own right to exist.

The values of woodlands are diverse, including

- their contribution to the natural beauty of the landscape,
- the rare species they may shelter,
- the erosion control and water conservation they provide,
- the oxygen they produce,
- the opportunities for recreation they provide, from hunting to nature study, from snowmobiling to cross-country skiing,
- their role as wildlife habitat,
- their role in conserving biodiversity,
- their ability to provide economic return through maple-syrup or firewood production, and
- their ability to provide a sustainable harvest of timber.

This wide range of values, which can be reflected in our decision making, is the other half of the new emphasis on holistic forestry – that is, the importance of taking all values into account when making management decisions.

Holistic forestry, or woodland stewardship as we use the term, is therefore forest management that considers the whole woodland as an ecosystem and all the management options that can be applied, as well as considering the whole range of values that the woodland provides.

Understanding Your Woodland

The first step in choosing among management options is to get to know your woodland. This requires walking through your woods at different times of the year, learning to identify the trees, and also recognizing other features such as streams or wetlands. There is no rush to make decisions; you will gain experience and be able to make better decisions over the years.

By gathering information about the trees, the other plants, and the wildlife in your woods, you can prepare a description or inventory. This is the first step in developing a management plan. It can also be the start of a much deeper appreciation of your woodland as an ecosystem in all its complexity.

You can either prepare such a woodland inventory yourself or hire a forest consultant to prepare one for you. Most government programs that support woodland management require the preparation of a basic inventory as the first part of a management plan. In Chapters 3 and 4, we outline the steps to follow in preparing a woodland inventory.

Balancing Choices

There are two fundamental choices for landowners to make among many possible woodland-management options and values.

1. The first choice is to determine a basic level of environmentally sustainable care for your woodland that will ensure ongoing protection of the entire woodland ecosystem. All woodland management, no matter what your personal objectives, should meet basic environmental goals.

2. The second choice is the balance of emphasis you wish to place among different values or uses reflecting your own personal interests, from conserving biodiversity and watching wildlife, to harvesting timber and firewood.

In this book, we will deal with both of these aspects of managing your woodlot.

In the first case, there are basic steps you should take to ensure the minimum level of environmental sustainability for your woodland. These include protection of the drainage pattern, protection of nesting birds from disturbance, and if harvesting timber, strict adherence to sustainable forest practices, among other concerns.

These basic requirements for sustainable woodlot management are emphasized throughout the book. They will be reflected in different chapters and depend on your own management interests. Remember, the best environmental management of your woodland is usually the best economic choice as well, especially in the long run.

Regarding the second choice, you may be a landowner whose primary interest is in simply appreciating the natural beauty of your woods and sometimes walking through it. You may want your woodland to contribute to the conservation of biodiversity in your region. On the other hand, you may wish to manage your woodland for sustainable firewood or timber harvest.

In Chapter 5, we discuss a range of management choices that emphasize environmental sustainability, support nature appreciation, and conserve biodiversity and wildlife habitat. In Chapter 6, we present a detailed discussion of sustainable management practices for firewood and timber harvesting.

Beyond these basic management choices, you may also be interested in reforestation. Whether planting pine and spruce plantations, planting hardwoods directly, or by allowing old fields to regrow into forest, many landowners are expanding their forest holdings. If your interest is in conservation of biodiversity, planting trees either to expand your woodland or to connect it with other woodland patches may be a high priority.

But few people who plant trees realize just what they are getting into. Coniferous plantations require thinning several times over their lifetimes. Hardwood plantings require lots of tender loving care to get well started. There are important benefits to using genetically appropriate local planting stock, and significant potential problems if you do not.

The natural regrowth of trees, a process we describe as natural "succession," will also create a new woodland quickly, but careful management can enhance this process. In Chapter 7, we review all these management options for reforestation.

Other more specific management options, from growing Christmas trees to making maple syrup, may be of interest to the woodlot owner. Often incorporated in farm operations as a source of extra income, many of these opportunities are described as "agroforestry" options. These are presented in Chapter 8.

Access to your woodland by trails or woodland roads is an important aspect of management, and a potential source of major environmental impact. Pests, whether in the form of disease, insects, or the two-legged variety, can be a major concern. In Chapter 9, we review the design and building of woodland trails and their benefits for both nature appreciation and timber harvesting, and we review some guidelines for coping with the pests that may invade your woodlot or plantation.

In Chapter 10, we discuss how all these ideas can be pulled together into a management plan for your woodland. There are significant benefits to be gained by establishing a written management plan, which you can use to organize your work, share with your heirs, or discuss with a professional forester.

If you are a hopeful woodlot owner looking to purchase rural woodland, you will find information to keep in mind presented in Chapter 11.

In Chapter 12, we present a number of options for ensuring the long-term future of your woodland. Through your management plan, a conservation easement, or through other legal mechanisms, you may gain financially while at the same time ensuring that your woodland remains well managed for decades to come.

After the practical aspects of woodland management have been dealt with, there remains the spirit of the woods. Chapter 13 includes some seasonal observations of our own woodlands over the year.

Finally, we provide an appendix of sources for further information: additional readings and contact addresses for professional assistance. We encourage you to take advantage of both for more specific help than is within the scope of this book. But with this book and some local professional assistance, we believe that you can acquire all you need to know to actively manage your own woodlot.

Seeking Professional Assistance

We believe that interested landowners can learn to understand and undertake most of the useful management in their woodlots if they wish. Certainly, landowners can learn to understand all the management options open to them and make meaningful choices in caring for their woodlands.

But there are many situations in which professional assistance can be invaluable; for example, when planning a timber harvest, identifying rare species, or when reforesting land.

Local professionals can also direct you to other information or contacts: government assistance programs, wildlife management advice, timber marketing, or purchasing trees for reforestation.

We encourage you to seek local professional assistance to complement the information you gain from this book. There is far too much regional variation and too many details at the individual property level for us to cover all local possibilities here. The exact growing conditions, the species present, the local product markets, and special government support programs all vary across this region. Our experience is in the northern part of this region; adjustments may be required to apply management principles to the different species mixes in the southern part of the eastern deciduous forest. A local expert can give you the advice you need in accessing this information.

We hope that this book takes you a long way toward your own woodlot-management plan, but do not hesitate to consult the experts in your own community. You will find them helpful, knowledgeable, pleasant to work with, and worth every penny of their time. Indeed, a friendly partnership with a local professional can be one of the most important supports you can have in clarifying details, providing other contacts and information when needed, and in giving you the confidence to go ahead with your woodlot-management decisions.

See the discussion on working with professionals in the appendix for further guidance and initial contact addresses for your state or province.

CHAPTER 2

WOODLAND ECOLOGY

"Ecology" is the science that studies the relationships of living organisms to each other and to their environments. It derives originally from the Greek work *oikos,* or household. In a very real sense, ecology looks at the earth as the home of all species, including humanity.

When studying ecology, we quickly come also to the term "ecosystem." An ecosystem is all the organisms in a given area, interacting with each other and with their environment. This environment includes many physical factors – such as soil and water, topography, drainage patterns, climate, and sunlight – that influence the lives of the organisms living there.

The interaction of these ecosystem components results in a complex array of cycles of material and energy. In any woodland, there are nutrient cycles of carbon and nitrogen, a water or hydrological cycle, and an energy cycle or food web – and these are only the most obvious relationships.

The interactions of the individual species with other species and with all these physical factors make the ecology of the woodland extremely complex. We do not usually see many aspects of the forest ecosystem, especially the thousands of fungi and

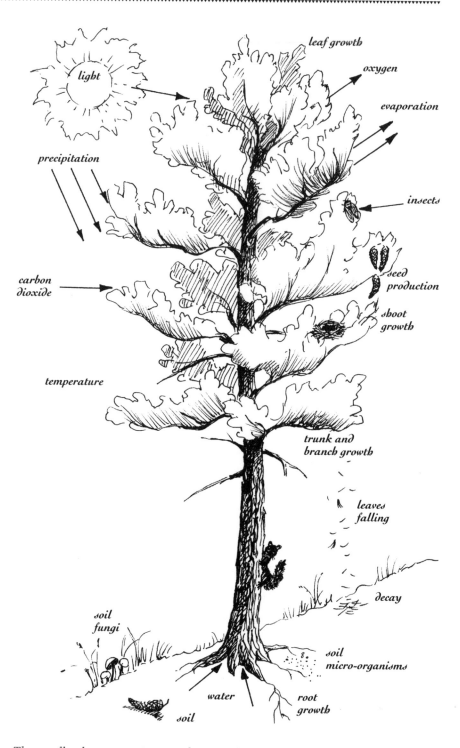

The woodland ecosystem is a complex mix of interactions, represented in the growth of a single tree, influenced by climate, water, and wildlife.

microorganisms that live in the soil, though they play critical roles in forest decay and tree growth.

All these relationships also change over time as trees and other plants in a forest grow and mature. Trees themselves change the environment for other plants and for wildlife; for example, by growing large enough to provide shade. This pattern of change is described as "natural succession," for the succession of changes that occur in the woodland ecosystem over time.

Landscape Influences

In understanding woodland ecology, you must also look at the broader landscape beyond the woodland for other important ecological relationships.

The term "ecosystem" is often replaced with the term "ecological community" or "vegetation community" when describing a local region. There may be many ecological communities in a broad landscape. There are ecological relationships among these communities as well as within them.

The edges of woodlands provide what biologists describe as "edge habitats," where two different communities meet. Many species of wildlife depend on more than one community for food or cover, so edge habitats tend to be attractive to them.

On the other hand, some species need the deep, undisturbed habitat of the forest interior for their life cycles. Of course, larger woodlands provide more interior habitats for these species and so are of critical importance.

Connections between woodlands are also an important ecological link. Birds, small mammals, and even plants (their seeds carried by the birds and small mammals), migrate along corridors such as fence rows or stream-valley forests.

In this chapter, we will first discuss the environmental factors that are important influences on woodland ecosystems. Secondly, we will discuss the many interrelationships that exist between species and their environments. Thirdly, we will look at the patterns of change in these relationships over time, which is known as natural succession. Finally, we will examine some of the ecological relationships in the broader landscape. Along the way, we will identify places where an understanding of woodland ecology has direct implications for woodland management. The principles of woodlot management are, in fact, directly based on an understanding of woodland ecology.

Soil and Nutrients

The first factor that makes up the physical environment of a woodland is soil. Soil provides direct structural support for growing plants, but also provides the needed nutrients, such as nitrogen, phosphorus, potassium, magnesium, and calcium. The continuous cycling of these nutrients through growth, decay, and reuse enables ongoing growth of the woodland.

Soil itself develops through the slow weathering of the underlying geological deposits, enriched by the addition of organic material, such as leaves. Though we may not see them, soils are teeming with microorganisms, many of which are decomposers, vital links in these nutrient cycles.

Soil types have a major effect on the plants that grow in a woodland. At one extreme is the rich, black organic muck of a wetland. Only trees and other plants adapted to the wet growing conditions of these soils will survive here. Drier soils range from sand and gravel through loam to clay.

Soils that are very coarse in texture, such as sands and gravels, have large spaces between the individual mineral particles. Thus water runs rapidly through such soils, and they tend to dry out quickly. They also tend to be poor in nutrients, since nutrients attach themselves to the smaller particles in the soil. Soils made up of very fine particles, such as clay and silt, are packed together densely with very few spaces between the mineral particles. Thus water cannot run quickly, if at all, through these soils, and they tend to be very wet, though they often have high nutrient levels.

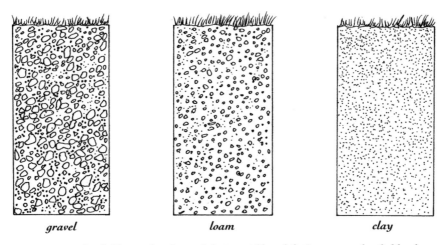

gravel   *loam*   *clay*

Coarse textured soils like sand and gravel drain quickly, while fine textured soils like clay drain very slowly. Drainage in turn is a major influence on tree growth. Medium-textured loam is best for most tree growth.

The ideal soil for forest growth, just as in a garden, is a rich loam. Loam is made up of a mixture of particle sizes, large enough to drain well, but fine enough to maintain good nutrient quality. Over time, such loams also become enriched with organic matter, further enhancing plant growth.

In a woodland, the best growth will be in well-drained, medium-textured loam soils. Very dry soils or very wet soils provide obvious limitations to tree growth. Soil depth can also be a critical influence. In some parts of the northeast, we find bedrock at the surface, with little or no soil at all. These extreme conditions are an obvious limitation for plant growth, but in fact, some types of bedrock, such as limestone, support other unique plant communities.

Individual species of plants, including trees, are all adapted to specific growing conditions. Sugar maple prefers a well-drained loam soil, but black ash will grow in wet organic soils. Thus the first influence on the woodland ecosystem is the mixture of soil types present.

MANAGEMENT IMPLICATIONS: Soil types will determine the productivity of existing woodlands and will influence the species you should choose for reforestation.

Topography

"Topography," or the actual shape of the physical landscape, is another important ecological influence. Slopes may be steep or almost flat, and they may face any direction. The direction in which a slope faces is called its "aspect." Altitude is also an important influencing factor.

The hardwood-forest region of eastern North America has an enormous range of topographic conditions, from areas near sea level to the northern Appalachian Mountains, from flat glacial lake beds to steep rocky cliffs. In between are thousands of square miles of rolling hills and river valleys.

Perhaps most unusual are the plant communities that are found in unique topographic locations. The best example is the ancient cedar forest on the vertical limestone cliffs of the Niagara Escarpment. Extending through New York State west to form Niagara Falls, then extending northwest through Ontario, and curving around the north side of Lake Michigan into Wisconsin, the cliffs of the Niagara Escarpment provide a unique habitat. The gnarled and twisted eastern white cedar that grow directly out of crevices in these cliffs have been shown to be up to a thousand years old, by far the oldest trees in eastern North America. Sometimes these trees actually hang down the cliffs from the ledge where their roots have gained a foothold.

Much more commonly, forest communities are influenced by the aspect of the slope on which they grow. Slopes facing south and west tend to receive more sunlight and be warmer, while north- and east-facing slopes are cooler and receive less sunlight. This may result in significant differences between ecological communities. A good example of this local temperature difference is the fact that all major ski hills in this region of the continent are located on north- or east-facing slopes, where the snow is less likely to melt in the direct sunlight.

Large slopes cause air to drain downwards; the movement of air keeps overnight temperatures on slopes slightly warmer than either

Ancient cedars growing along the Niagara Escarpment are the old-est trees in eastern North America with some living to 1,000 years.

at the top or bottom reducing the risk of frost. Thus fruit trees are often grown on slopes to minimize late spring frost danger.

Altitude is another topographic influence. This is most notice-able in mountainous regions. Average temperature falls as you go upwards in altitude, until you reach ice and snow if the mountains are high enough. This changing temperature results in different vegetation communities at different altitudes.

MANAGEMENT IMPLICATIONS: The species of vegetation growing in your woodlot, as well as your options for tim-ber harvesting or trail building, may be limited by topo-graphic location and slope.

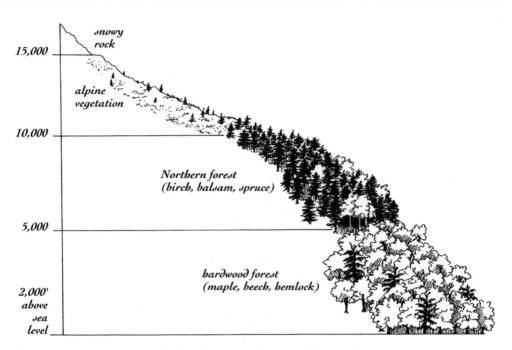

snowy
rock

15,000

alpine
vegetation

10,000

Northern forest
(birch, balsam, spruce)

5,000

hardwood forest
(maple, beech, hemlock)

2,000'
above
sea
level

Forest composition varies with altitude. As temperature declines with rising elevation, the species which can survive define a changing forest community.

Micro-Topography

Within a woodlot there may be small variations in topography: steep slopes, depressions that form wetlands, and drier upland areas. One of the most interesting is the "micro-topography" of pits and mounds.

When large trees are uprooted in a woodland, a massed tangle of roots is thrust into the air, bringing with it organic debris and mineral soil. Below, the mineral soil is often exposed in the hollow left by the uprooted tree. Over time, the roots will rot and subside as they are covered with fallen leaves and organic matter, gradually forming a mound. At the same time, the hole left by the roots will gradually modify into a smoother depression or pit. This is the pits-and-mounds topography of an undisturbed old forest.

Along with pits and mounds are a variety of biological life. The Louisiana waterthrush, a rare woodland-dwelling bird, depends on root tangles for nesting sites, especially if the wet pits fill with water. On the other hand, the exposed mineral soil in the pits often provides a drier soil in which oaks, white pine, and basswood can germinate. Salamanders, one of the largest but often unseen biological components of many woodlands, live beneath and eventually within the fallen log. Invertebrates, bacteria, and fungi contribute to the process of decay.

Together, these micro-topography variations add immensely to the biological diversity of a woodland.

Water

Water is the third critical element of the woodland ecosystem, serving as a vital connector between the soil and its nutrients, and plant growth. In this way, water is essential for all life on earth. It also shapes the physical landscape, forming unique features such as streams, springs, and wetlands.

Water arrives in the form of precipitation, either as rain or snow, and percolates through the soil to become groundwater, as well as running off into streams. Water also makes up a significant portion of the cells of living plants. Along with carbon dioxide from the atmosphere, water is one of the basic ingredients in photosynthesis, in which the plant converts raw materials into sugar in the form of glucose.

This role of water in the ecosystem is referred to as the "hydrological cycle." Water falls and runs off the surface or infiltrates the ground. Eventually it will reach a body of water, such as the ocean, from which it evaporates, starting the cycle again. Along the way, many molecules of water are intercepted by plants. Water enters the roots and leaves, travels through the plant, and is either used in photosynthesis or transpired back into the atmosphere as the plant breathes.

Water may also cause erosion on steep slopes as it acts on the landscape. Protected by a canopy of trees, the slope may be stable, but when clear-cut, a slope is highly susceptible to severe erosion. Similarly, vegetation along the banks of a stream or surrounding a

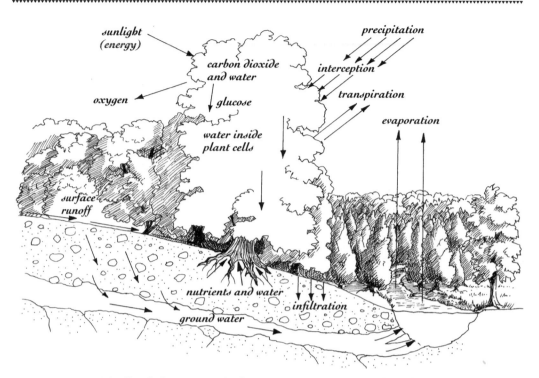

sunlight (energy)

precipitation

carbon dioxide and water

interception

oxygen

glucose

transpiration

evaporation

water inside plant cells

surface runoff

nutrients and water

infiltration

ground water

Woodland plays a critical role in maintaining a natural hydrological cycle intercepting rainfall and slowing down runoff.

wetland prevents direct runoff, holding back sediments or fertilizer that may otherwise be carried into the channel.

On open agricultural fields, or even more so in urban areas, water runs off quickly into streams and rivers. In the woodland, water is intercepted and used by trees directly, or runoff is slowed so water infiltrates the soil replenishing the groundwater. Groundwater, in turn, replenishes ponds, wetlands, and streams. With sufficient groundwater, springs develop, smaller streams run all year, ponds and wetlands don't dry out, and water quality is higher.

A "watershed," or drainage basin, is the area of land drained by one particular stream or river system. Within any large watershed there are many smaller watersheds, or sub-watersheds, representing the drainage area of smaller streams.

In two comparable watersheds – one forested and the other cleared for agriculture – streams in the forested landscape will have higher water levels, higher water quality, steadier flow patterns over the season, and healthier fish populations. Woodlands play a critical role in maintaining a healthy hydrological system.

> MANAGEMENT IMPLICATIONS: The relationship between soil conditions and water is an important one to look for in planning reforestation. To ensure healthy growth, species for planting must be chosen to match these environmental conditions. Drainage features should be protected carefully during timber harvesting or trail building.

Climate and Sunlight

Climate is the large pattern of typical weather that characterizes a region. Temperature and precipitation vary throughout the year. Wind may provide occasional damaging storms, and of course, sunlight is the engine that drives the entire woodland ecosystem, just as it provides the basic energy source for all other life on earth.

The climate of eastern North America is characterized by varying seasons as temperatures rise and fall. The seasons are one of the most noticeable features of woodlands, due to the brilliant colors of the leaves, especially those of sugar maple, as the trees shift from their summer growing season to winter dormancy. Precipitation also varies with the seasons, arriving often in the form of thunderstorms during the summer and more commonly as passing weather fronts during the rest of the year.

In fact, the seasons are so obvious in eastern North America that we may take them for granted. They define our relationship to woodlands so intuitively in the activities we undertake, from planting trees in the spring to cutting firewood in the fall. Plants and wildlife have had to develop an amazing range of adaptations in order to survive such a cycle of changes over the year.

Deciduous trees lose their leaves and shut down the photosynthesis process over the winter; coniferous trees have adapted by having leaves in the form of narrow needles, partly to assist in winter survival. Animals hibernate, migrate, or find shelter during winter seasons. Patterns of both hibernation and migration reveal surprising adaptability. Some frogs and toads, buried in mud over the winter, can survive after having portions of their body frozen. Birds fly thousands of miles to winter in the tropics.

Our climate also provides a reservoir of atmospheric gases, especially carbon dixoide, that are critical to plant growth and development, and through this growth, plants contribute gases,

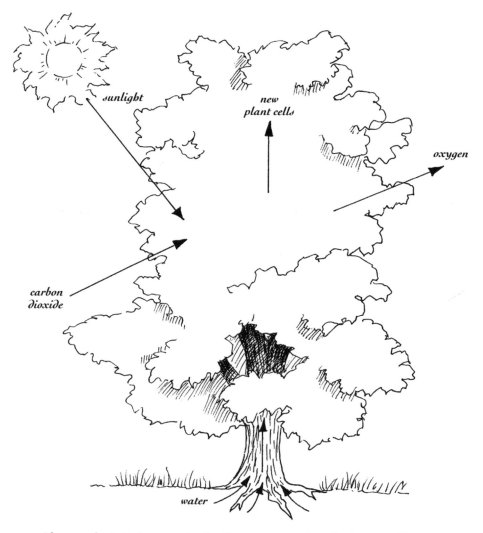

Photosynthesis is the true miracle of creation, providing the basis of all life on earth by combining carbon dioxide and water in the presence of energy from the sun to create carbon molecules and oxygen.

especially oxygen, back into the atmosphere. Summer sunlight is the vital key that powers this living system.

The component that changes the soil, nutrients, water, and atmospheric gases – all inherently non-living parts of the physical environment – into a living system is sunlight. Sunlight provides light energy, which is the foundation of all life on earth. Plants contain chlorophyll, which captures this light energy and changes it into chemical energy.

This occurs through the process of "photosynthesis," meaning "put together with light." Carbon dioxide and water are combined to form glucose, a simple sugar; oxygen is given off as a by-product, contributing to the balance of gases in our atmosphere. The energy is held in the carbon bonds of the sugar molecule. These sugars eventually form the living structure of the plant.

Animals (and people) also depend on these sugars; all food we eat ultimately derives from photosynthesis in plants. This is the way we capture the light energy of the sun.

An important concept in woodland ecology is the difference between "tolerant" and "intolerant" tree species. Trees vary in the efficiency with which they can intercept and use the sun's light. Trees with greater efficiency, such as beech and maple, can use

Table 1   Tolerant and intolerant tree species

SPECIES TOLERANT OF SHADING	INTERMEDIATE SPECIES	SPECIES INTOLERANT OF SHADING
Balsam fir	Black ash	Black cherry
Beech	Elm	Black walnut
Eastern hemlock	Red maple	Hickory
Hop hornbeam	Red oak	Red pine
Silver maple	White ash	Tamarack
Sugar maple	White pine	Trembling aspen
White cedar	Yellow birch	White birch
White spruce		

Sources: T.J. McEvoy. *Introduction to Forest Ecology and Silviculture.* Burlington, VT: University of Vermont, 1990.
S.N. Staley. *Wood: Take a Stand and Make It Better.* Toronto: Ministry of Natural Resources, 1991.

even the diffuse, lower-light levels of the forest interior for photo-synthesis. These trees are termed "tolerant" because they are tolerant of such shaded conditions. Trees that require bright direct sunlight are termed "intolerant" because they are intolerant of shaded conditions.

> MANAGEMENT IMPLICATIONS: Regrowth of trees after thinning or harvesting is heavily influenced by light conditions. Minimal thinning favors shade-tolerant species such as sugar maple, while extensive thinning or open-patch cuts favor more intolerant species such as ash, oak, or cherry, provided there is a seed source nearby.

Competition

All plants in the woods compete for nutrients, water, and sunlight. In a dense stand of trees, individual trees will have to compete for these resources, and will not grow as fast as they otherwise would. Trees growing in the open may have far less competition and, therefore, grow faster. They will also develop large spreading crowns, ideal for maple syrup production. In fact, such open-grown maples produce sap with a much higher sugar content; their leaf-filled crowns are a sugar factory.

Inside the woods, trees competing for sunlight reach toward the forest canopy to capture their share of the sun's energy. In these conditions, trees grow tall and straight; lower branches die and fall off, leaving straight trunks that make ideal sawlogs for timber harvesting. It is easy to understand that if such trees are too crowded, each one will have a smaller crown, therefore producing less total energy in a year and less growth. At the other extreme, with no competition trees develop many lower branches and a wider growth form, leaving the trunk unuseable for timber.

This phenomenon has led foresters to develop a concept called "stocking." In theory, there is an ideal number of trees where

Trees compete for sunlight and nutrients, growing more slowly if crowded and growing larger more quickly with less competition.

growth is maximized because trees are not too crowded, but they are still crowded enough to produce trunks that are tall and straight. This ideal number of trees is the "stocking rate", a practice used to optimize growth for timber production values.

> MANAGEMENT IMPLICATIONS: Active management of a woodlot for timber production aims to calculate a stocking rate, and plans thinning and harvesting operations to maintain a position close to this optimum for best long-term growth.

There are many other examples of competition in the woodland ecosystem besides trees. Spring wildflowers gain their edge in the competition for sunlight by growing and blooming before the leaves come out on the trees. Many also have large leaves, ideal for intercepting the low light levels of the forest interior.

Competition is even more important in reforestation. Young tree seedlings have a difficult time competing with the other vegetation in an open old-field ecosystem. Grasses and other plants efficiently capture the sunlight, nutrients, and moisture through rapid growth in the spring. Deciduous trees, in particular, need to be protected from such competition if they are to survive, let alone grow quickly. In fact, this is the reason that conifers such as pine and spruce are used so frequently for reforestation; they can survive such competition, and though they grow slowly until they exceed the height of the surrounding grasses, they do not need the care required by deciduous seedlings.

> MANAGEMENT IMPLICATIONS: Deciduous tree seedlings need good protection from the competition of surrounding grasses and other vegetation to survive. Mulches, tree shelters, or spraying are used to provide this. These protection practices can also be used on conifers to enhance their growth.

The Soil Ecosystem

Many of the most fascinating relationships in the woodland ecosystem go on unseen underneath your feet. Woodland soil is an ecological community in itself, teeming with millions of microorganisms and fungi that are essential to continued life. These organisms decompose fallen leaves and other material, enabling the continuation of nutrient cycles that sustain life in the forest.

Many people are surprised to learn that in a typical woodlot there is more biomass in the soil under our feet than above ground, and there are many more species of organisms.

A critical link in the ecosystem is provided by what are known as "mycorrhizal" fungi or "mycorrhizae." The fungi known as mycorrhizae live on the roots of trees such as sugar maple in a mutually beneficial relationship, called a "symbiotic" relationship. The fungi help the tree gain nutrients and water from the soil by increasing the effective root-surface area; in turn, the fungi gain some of the sugars produced by the tree.

Wildlife Habitats

Above the surface of the soil, there are similar interactions between the plants in the woodland and the wildlife that lives there. Many relationships are "symbiotic," or mutually beneficial. Squirrels eat acorns, but they bury a few too, generating more oak trees. Birds eat cherries, but spread these seeds as well. To survive, wildlife requires food, safe cover for nesting and shelter, as well as water. Woodlands provide all these requirements in many different

Wildlife need food, nesting cover, water, and space for their habitat.

forms. Not only the fruit of trees and other plants, but the bark and leaves provide food. Trees, hollow logs, holes in dead standing trees or "snags," dens in the ground, and logs on the forest floor all provide cover for nesting.

Woodpeckers are an example of a key species from which many other species benefit. Excavating their nesting cavities in hollow trees, woodpeckers create thousands of cavities for other birds to use later. Dozens of bird species, from tiny warblers to wood ducks, depend on the original cavities chipped out by woodpeckers for their own nesting cover.

Cavities created by woodpeckers later provide homes for dozens of other species.

But some relationships can be damaging. Porcupines kill trees by girdling the bark; insect infestations can be far more devastating. Wildlife competes with other wildlife as well; chipmunks, red squirrels, and aggressive birds such as blue jays eat the eggs and young of other birds.

Food Chains and Energy Flow

All these interactions between plants and wildlife form a food chain, or more correctly a food web, which provides energy flow through the forest ecosystem. Plants are the direct producers of useable energy through photosynthesis. They provide food for many other organisms, from monarch butterflies and chickadees to white-tailed deer.

In turn, other wildlife prey on these species. Birds eat butterflies, and then are eaten by larger birds. Foxes feast on mice and

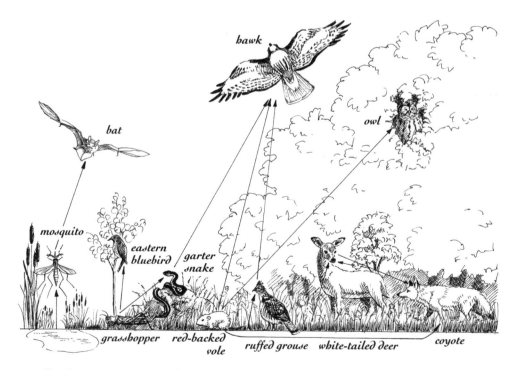

Woodland ecosystems form food webs in which plants, insects, mammals and birds are all eaten, and eventually die and decay, forming an energy cycle.

other small mammals. Coyotes or larger carnivores kill deer. At the top of this pyramid of consumers are people, who eat both plants and meat.

At all levels of this food web are decomposing organisms, which assist in the process of decay of any organism that dies. And so the entire food web becomes a cycle, returning the nutrients to the ecosystem once again. Through the cycle, the energy generated originally by photosynthesis is transferred from one organism to another. The entire functioning ecosystem depends on the original carbon dioxide, oxygen, water, and sunlight.

In fact, trees are quite amazing organisms in this complex biological system. Through the mechanism of photosynthesis they produce most of the energy – in the form of sugars or glucose – that the entire woodland ecosystem needs to survive. At the same time, they build their own structure of wood cells, creating a physical body that may reach 100 feet (30 m) in height – strong, yet amazingly resilient in a wind. In turn, these trees provide a habitat, enriching organic matter, food, and shelter, all while experiencing a seasonal cycle of growth finely attuned to the climate and landscape in which they live.

> MANAGEMENT IMPLICATIONS: Understanding the components of wildlife habitats and the relationship between species leads to specific wildlife-management options, such a building brush piles for shelter or maintaining tree species as food sources.

Patterns of Change

A woodland is constantly changing. Much of the change you cannot see. Slowly trees get larger and taller; plants in the understory grow and die. Wildlife moves constantly in search of food or shelter. Water moves through the landscape, either in surface streams or as groundwater. Air moves through the trees creating a breeze.

Inside the plants and soil, other action is going on. In plants, water, nutrients, and gases are constantly moving up and down; the process of photosynthesis comes and goes with the light. Leaves on the plants breathe much as we do. In the soil, microorganisms are always on the move, eating and aiding in decay.

A woodland ecosystem is not merely what you see at one point in time; it is a living, complex process of energy creation and use on a minute-by-minute, day-by-day basis.

Over much longer periods of time, the change in forest composition can be more readily seen. Old fields, once used for agriculture, gradually change as grasses, small shrubs, and young trees invade them. Trees grow, and new species appear. Gradually, through several sequences, the forest changes until a mature forest results.

This process of growth and change over time is known as "natural succession." Individual species succeed one another as part of this process through natural ecological cycles of germination, growth, and death.

The patterns of natural succession in the forests of eastern North America are well known. The appearance of individual species is controlled largely by sunlight. Species that can grow in open, bright sunlight appear first. Known as pioneer species, these include grasses and many plants we commonly know as weeds,

As old fields grow into forest, grasses and weeds are replaced by shrubs, sun-loving trees, and finally trees which tolerate shady conditions. This pattern is known as natural succession.

such as asters and goldenrods. If a dense cover of these species is established, this old-field stage may persist for many years, but as it grows, it changes the conditions for other plants, and new species appear.

Often the second stage of natural succession involves the spread of shrubs such as dogwoods, wild grape, or raspberries. In wet conditions, young willow may appear. At this stage, the canopy is a little higher, and in the shade of shrubs some of the grasses or other plants may die, leaving room for young trees to germinate. Soon the first tree species will appear.

Typically the new species are those that still require some open sunlight for growth, or are intolerant of shade. They are often described as "early successional species." Exactly which species appear depends on available seed sources. Downwind from a mature ash tree in a fencerow, for example, hundreds of young ash may appear quickly in a field. Among the typical early successional species are white birch and various poplars, especially trembling and large-toothed aspen. Planted pines, such as red and white pine, often do well in this stage as well.

As this second group of trees matures, a forest canopy begins to form, and underneath it still other new species appear. At this "mid-successional stage," species such as red and white oak, white cedar, and red or white spruce appear. The vegetation community now begins to resemble a woodland, though the successional process is not yet finished.

As deeper shade develops, the shade-tolerant species, such as sugar maple, beech, and hemlock, can take hold. These trees would not survive well in open sunlight, but will gradually become dominant given a long enough period of time under a forest canopy of shade because of their shade tolerance.

Of course, many other variables exist as this sequence of natural succession occurs. Light levels, nutrient-cycling patterns, and soil structure all change. As well, wildlife populations change dramatically. In the old-field ecosystem, meadowlarks, rabbits, and field mice may be common, but as succession develops, ruffed grouse, white-tailed deer, song sparrows, and juncos appear. By the time a mature forest is in place, its wildlife includes red squirrels, porcupine, veery, and downy woodpeckers, while species that like the open fields will no longer be found there.

As well, there are larger patterns of change in the landscapes of eastern North America as a whole. In areas of more rugged topography or shallow, unproductive soils, farms cleared during settlement in the last century are now reverting to forest through natural succession. In some areas of New England and adjacent parts of Canada, there is now considerably more forest cover than was present thirty years ago, and this trend is continuing. Often this is associated with rural non-farm residents buying country property and commuting, or else retiring to rural areas. It may lead to a gradual revival of local forest industries in the future.

At the other extreme, particularly on the western edge of the deciduous forest region on high-quality agricultural soils, farming is intensifying, and remaining woodlots are increasingly fragmented into very small parcels or cleared entirely. Here the need for woodland conservation to conserve biodiversity is critical.

Woodlands in the Broader Landscape

Not only is your own woodland a fascinating, changing ecosystem, but it is part of a bigger ecosystem: the landscape surrounding it. And that landscape is part of a still bigger one: the entire eastern North American region.

As well as looking at your woodland to understand all the ecological processes affecting it, you should see it as part of this broader landscape.

For example, if your woods is the only woodland in an open agriculture landscape, it will be a very important habitat, regardless of its size. On the other hand, if your woodland occurs in the midst of a larger forested region and also has a small area of meadow, the meadow may be the unusual habitat especially worth protecting.

In some parts of eastern North America, the patchwork land-scape of small farm fields and woodlots is gradually disappearing as farms revert to forest and larger areas of continuous woodland become more common. In these regions, management to maintain old-field or meadow habitats may be worthwhile.

In other parts of eastern North America, most of the remaining woodland has been fragmented into small patches due to the spread of agriculture and urbanization. This has changed the forest ecosystem dramatically. The original continuous forest had exten-sive interior-forest habitat; today the many small patches of forest mean there is dramatically more edge habitat. As a consequence, species that depend on the interior-forest habitat have experienced severe declines in numbers, while species that enjoy edge habitats have increased significantly.

This is a serious conservation issue, as some species requiring the habitat of the forest interior are declining over large parts of their range.

In fact, in the past, much effort has been made in wildlife man-agement to increase edge habitat, because this provides popula-

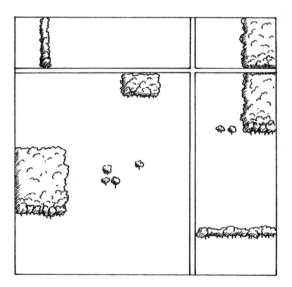

The fragmented forest remnants of an agricultural landscape leave little habitat for species that only live in the forest interior.

tions of animals for hunting, such as rabbit, partridge, and white-tailed deer. But with more recent recognition that all species are equally worthy of conservation, there is a new emphasis on conservation of biodiversity. It is in this context that the science of "landscape ecology" has emphasized the importance of large woodland patches and of connecting corridors between woodlands.

The Importance of Large Woodlands

Birds like to live and nest in different habitats. Some nest in open fields, some in wetlands, and some only nest in the forest interior. These forest-interior species need the shelter of surrounding woodlands for protection from predators.

The best example of such predators is the brown-headed cowbird, which lays its eggs in the nests of smaller birds, then abandons them, leaving the other birds to raise its young. Tiny warblers end up raising cowbird chicks bigger than themselves, and their own young perish in competition with the bigger cowbirds.

Cowbirds can fly some distance into a woodland from the forest edge to find nests to exploit, so nests near the edge are not protected. In fact, scientists suggest that nests are protected only if they are more than 200 to 300 yards (180–270 m) inside the woodland.

The forest-interior habitat exists only at the center of the woods, inside a zone up to 330 yards (300 m) wide all around the woods. Consequently, a woodland has to be more than 600 yards (550 m) across to have any actual interior habitat. It is not surprising that species that depend on these specific habitat conditions, such as the oven bird or the wood thrush, are in serious decline in North America. Many remaining woodlands in our landscape are simply not large enough to provide this habitat, though they do provide important habitats for many other species.

Large woodlands are, therefore, worth protecting where they exist, especially in landscapes where agriculture is more intensive, resulting in a very fragmented pattern of small woodlands.

The shape of woodlands also influences the amount of forest-interior habitat. A 60-acre (24 ha) woodland that is a solid block provides more interior area than a 60-acre (24 ha) irregularly shaped patch.

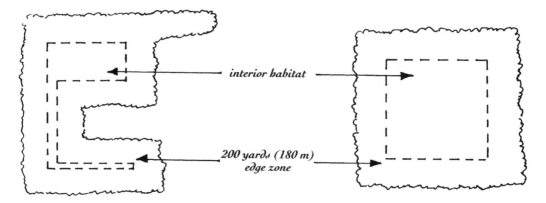

Shape is a major influence on the amount of "interior habitat" in a woodland; larger solid blocks provide significantly more interior space.

This is not to say that small woodlots are unimportant. Although large woodlots are a vital breeding habitat for several species of birds, plants are not so affected by woodlot size. Even a small woodlot can support a significant, varied vegetation community.

> MANAGEMENT IMPLICATIONS: Large woodlots are particularly important for conserving biodiversity. Enlarging your woodlot or filling it in to form a more solid block may be an important management choice.

The Importance of Natural Connections

Science has also revealed that corridors of natural vegetation help to maintain populations of wildlife (or biodiversity) in and between woodlands. In a small woodland patch, isolated in the middle of a city or farmland, some species can become locally extinct. Unless there is the opportunity for movement of species between wooded areas, new populations may not reestablish themselves.

Furthermore, without the movement of small mammals, genetic inbreeding may occur, leading to genetic impoverishment or local extinctions. Interaction among populations is critical to enable successful breeding and to maintain genetic diversity. Conservation of biodiversity includes measures to maintain genetic diversity within species, as well as the species themselves.

Larger mammals and birds move between woodlots, at least in rural areas, but vegetated corridors enable smaller species such as chipmunks to move more freely. Natural connections between habitats are therefore another important factor in the conservation of woodlands.

In this context, even small woodlots and narrow vegetation corridors play a significant role in maintaining healthy wildlife populations.

Streams and waterways provide an obvious place to locate such corridors of natural vegetation. Bands of natural vegetation along streams not only serve as wildlife corridors, but do double duty as buffers to protect the water quality of the stream. At the same time, these "riparian" (located at the water's edge) communities are one of the most biologically productive ecotypes in the landscape, supporting more diverse insect, bird, and mammal populations than anywhere else.

> MANAGEMENT IMPLICATIONS: Maintaining or restoring vegetation corridors, such as fencerows, between woodlots is important for conserving biodiversity. The wider the corridor the better, but any corridor is better than none; corridors along streams are particularly valuable.

Summary

Woodlands are complex systems of physical factors – such as soil, water, climate, and atmospheric gases – interacting with plants and animals in a dynamic whole, or ecosystem. Understanding woodland ecology requires looking at this big picture in all its complexity.

Ecology focuses on the relationships and cycles in a woodland: the relationship between water and plant growth, between the atmosphere and photosynthesis, and between wildlife and its habitat. Understood as an ecosystem, a woodland is a group of cycles: the energy cycle, the hydrological cycle, the carbon cycle, and others.

Growth in a woodland leads to changes over time – known as natural succession – in which both plant and animal species succeed one another in the ecosystem as changes occur.

Woodlands also have to be seen as components in the broader landscape if biodiversity is to be conserved. Today's science suggests the importance of both larger woodlands and natural connecting corridors between woodlands.

Now that you have some basic understanding of woodland ecology, we turn next to the work of preparing a woodland inventory.

Further Reading

Hunter, M.L. *Wildlife, Forests, and Forestry: Principles of Managing Forests for Biological Diversity*. New York: Prentice hall, 1980.

McEvoy, T.J. *Introduction to Forest Ecology and Silviculture*. Burlington, VT: University of Vermont, 1995.

Minckler, L.S. Woodland Ecology: *Environmental Forestry for the Small Owner*. Syracuse, NY: Syracuse University Press, 1980.

Patton, D.R. *Wildlife Habitat Relationships in Forested Ecosystems*. Portland, OR: Timber Press, 1997.

Richberger, W.E., and R.A. Howard. *Understanding Forest Ecosystems*. Ithaca, NY: Cornell University Extension Service, 1980.

Yahner, R.H. *Eastern Deciduous Forest: Ecology and Wildlife Conservation*. Minneapolis: University of Minnesota Press, 1995.

CHAPTER 3

Preparing a Woodland Inventory

The first step in deciding how to manage your woodland is to understand it as it exists today. Preparing a description or inventory of your woodland enables you to evaluate your choices and is the first step in developing a woodland-stewardship plan. Whether your interest is in nature appreciation or in timber harvesting, preparing an inventory will prove useful to you. Many government support programs for woodland owners are contingent upon the completion of a simple inventory.

You may choose to prepare your own woodland inventory and stewardship plan, or you may choose to hire a consultant to do this for you. In either case, this chapter will provide the background you need. It is our belief that most landowners can learn to do this themselves, but will require some professional assistance on specific details. By understanding the process of preparing an inventory, you can spend your money on professional consultants more wisely.

Most rural landowners use their woodlands for nature appreciation, such as watching wildlife, or for recreation; others require some economic return through timber harvesting. Many landowners want to cut a little firewood. Regardless of your interest, the most important management decision for your woodland is simply

to keep it. Maintaining your woodland is the most important conservation decision you can make.

Preparing a description or inventory of your woodland is the
basis for deciding on related management actions. Once you
understand the woodland ecosystem, you can usually conserve
your woodland for nature appreciation or for wildlife while also
enjoying some economic return. Good ecological management can
also be good economic management. Preparing an inventory will
help you understand your own woodland and then assist you in
choosing the most beneficial management options.

In this chapter we describe a basic inventory that will help you
regardless of your management interests. The inventory described
here will give you the information you need to make initial management choices no matter what your interests. You may then decide to
gather more detailed information on specific features such as rare
species, wildlife, or timber potential. In the next chapter we outline
in greater detail the additional information needed if you wish to
manage your woodlot for firewood and timber production.

This chapter is organized in a series of five steps that will help
you to compile your basic woodland inventory:

Step 1:   Walk through your woodland
Step 2:   Map the ecological communities
Step 3:   Identify the trees
Step 4:   Add information on cultural, physical and
          special biological features
Step 5:   Draw the final sketch map and summarize
          information

First, we review four basic definitions with which you will need
to be familiar.

Trees are commonly referred to as coniferous or deciduous, or
sometimes as softwoods and hardwoods. In Latin, the term "coniferous" means cone bearing and "deciduous" means falling off.

Conifers are classified as cone-bearing trees, although most
people probably recognize them more through their needle-like
foliage (although some, such as cedar, have scales rather than

needles). Conifers are also widely referred to as evergreens, though the larch is a notable exception in that it drops its needles each fall. In our region, the notable conifers, or coniferous tree species, include all pines, spruce, fir, cedar, and larch or tamarack.

Deciduous trees are those which drop their leaves in the fall and regrow new leaves every spring. Maples, oaks, ashes, hickories, poplars, and many more fall into this category. The eastern deciduous forest region is dominated by deciduous tree species.

In forestry terms, the coniferous species are often referred to as softwoods, while the deciduous species are commonly labeled hardwoods. In most cases, the wood of conifers is actually softer than that of deciduous species. Even more confusing, some foresters refer to sugar maple as hard maple, and red and silver maple as soft maples, though all maples are referred to as hardwoods. For our purposes we will try to stick to the terms deciduous and coniferous.

Step 1: Walk Through Your Woodland

Walk through your woodland and explore what you own, taking care and time to stop, look around, observe, and listen. Build your own mental picture of your woodland and begin to recognize special features. Go for walks in late April or early May and in September or October for the best views of your woods, but go during the winter, too, to watch for wildlife tracks. If you walk during mosquito season, go prepared.

Watch for four things during your initial woodland walks.

1. ECOLOGICAL COMMUNITIES: The most basic information needed in a woodland inventory is a list and sketch map of "ecological communities." Ecological communities are areas of your land where the trees, saplings, and other vegetation are relatively similar; for example,

- an area of older maple and beech trees in a mature forest,
- a lowland swamp composed of cedar trees,

Take a notepad, pen, paper, camera; whatever you are comfortable with as tools to record your observations.

- a field reforested with pine trees, or
- an old field now growing young ash and elm trees.

These ecological communities might also be described as vegetation communities, management compartments, forest units, or forest stands. Watch for visually different plant communities. Later you can come back to identify the actual species of trees present.

2. PHYSICAL FEATURES: You also need to describe the main physical differences or features on your land, including

- water and drainage patterns, inclusive of streams, ponds, and wetlands,
- steeply sloping areas,
- low-lying wet areas,

- high dry areas, and
- different major soil types or rock outcrops.

3. SPECIAL VEGETATION OR WILDLIFE: You might find several different types of biological features or wildlife habitats. These could include

- large hollow trees that provide dens for wildlife,
- a stand of spring wildflowers,
- brush or stone piles that provide homes for animals, or
- standing dead trees or "snags" that are home to many bird species.

You may see wildlife, or evidence of it, such as tracks, nests, droppings, or indications of feeding. Watch for birds and mammals, as well as reptiles and amphibians such as frogs and salamanders.

You may also discover rare species in your woodland. Though you may need assistance to be sure of this, it is well worth the trouble.

4. CULTURAL FEATURES: Finally, you need to describe any human-made features of the woodland, such as

- roads or trails, fences, buildings, and access points,
- garbage dumps or old foundations, or
- evidence that hunters use your woodland.

Neighbors may be able to provide some of this information and may also help you check the boundaries of your woodland, if you need to.

To gather information, it may be easiest to walk through your woodland at different times, and make four separate lists, one on each of the above categories. You may even be able to do this sitting at the kitchen table, if you are already familiar with your woodland.

Be Safe!

It pays to be prepared when walking in the woods.

- Dress for field work; wear appropriate clothes and boots.

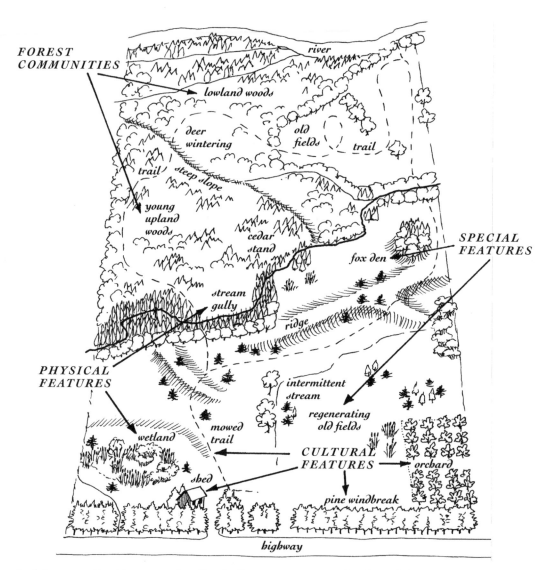

In doing a woodland inventory describe the forest communities, the physical features, any special vegetation or wildlife, and any cultural or human features.

- Take some bug repellent.
- If you have any doubts about getting lost, take along a compass. You can easily follow a straight line to get out of the woods if you need to.
- Experts doing regular field work also carry a loud whistle to call for help, and in inaccessible areas, a small first-aid kit.

- Don't forget a pen and notebook with which to take notes.

Watch out for poison ivy. Learn to recognize it and avoid it. If in doubt, avoid any plant with these characteristic three leaves.

Recognizing and Marking Boundaries

Before you spend time describing your woodland, you need to be sure that you know the boundaries. This is especially important if you are hiring a consultant to prepare your inventory (they will depend on you to show them the boundaries) or if you are planning to harvest timber (cutting your neighbor's trees is not a good idea).

If your woodland is an isolated woodlot within your farm, there may be no trouble at all in doing this step. You can see the woodlot among the farm fields, with all four sides clearly marking its boundaries.

On the other hand, if your woodland is continuous with others, if you own only part of a larger wooded area, then you may need to spend some time confirming your boundaries, especially if you are going to cut any trees.

You can do this in several ways:
- Walk the boundaries into the woodland and see if you can find a fence, even an old broken-down one, that marks property lines.
- Look up the survey of your land to find the correct shape and size of your property and the location of any survey markers. Find and mark the corner posts.
- Buy an aerial photo or a detailed topographic map to

establish the size and shape of the property that you
own (more explanation of this below).

- Knowing your own property boundaries and location,
  you can follow a compass bearing into the wooded area
  to establish a boundary. In addition, you can pace out a
  certain distance to help establish boundary points.
  (Pacing is described on page 54.)
- If you do not have a fence line to mark your boundaries, you may need to use some flagging tape or
  another obvious marker. You can buy brightly colored
  flagging tape at most hardware stores.

Working with your neighbors is a good move at this point. Older
residents may know where survey stakes are, and it's much easier to
agree on property lines before you make a mistake than after.

Step 2: Map the Ecological Communities

The basic inventory needed for woodland stewardship is just
a description of each major ecological community, along with
physical, cultural, and other special features that make up the
entire woodland.

These communities are important because management
options will likely be different in each major community. For example, you would manage a reforested pine plantation with different
techniques than you would use in a mature hardwood forest. You
may manage a young stand of deciduous trees in different ways
than a young stand of coniferous trees.

Among all the ecological characteristics of your woodland, the
vegetation communities are one of the easiest to recognize and
describe. They provide the simplest framework to start understanding the biodiversity in your woodland. To prepare your map,
use the following guidelines:

- review your walk through the woods, and decide how
  many major vegetation communities there are (don't

bother with tiny little patches of vegetation that are different, just look for major differences; separate communities should be at least 1 or 2 acres [1 ha] in size, usually much larger);

- sketch out the basic shape of each community as part of your entire woodland.

Look for differences in both vegetation and in physical conditions; areas that are wet and those that are dry will usually support different forest communities. You can also look for special features as you go; later you will come back to make a list of these.

Using Aerial Photographs

It may help you to bring along an aerial photo of your land when you take your walk, making it easier to locate and later to map the features you locate. It may also show you patterns on your land that you were unaware of.

Aerial photographs are especially helpful when walking your land and drawing a sketch map, since they are actual photographs of the landscape. Features on the ground, such as roads, buildings, and areas of forest vegetation, show up on the photo and help you to easily locate your woodlot. Aerial photos are also useful to give you an idea of the shape of the property you are studying or to help you establish the size of your woodland.

Aerial photos are usually available from your provincial or state government. In our region, the most recent standard aerial photos available are black-and-white photos taken in 1978 at a scale of 1:10,000. The year and scale of photos varies depending on the purpose and location of the original photography, so be sure you know the scale of photos you are using.

Remember that maps and aerial photos are often ten to twenty years old, and land uses may have changed in the surrounding area. Always check information in aerial photos and maps against what you find on the ground.

It is usually very easy to understand an aerial photo because ordinary features are visible on it. The shape of roads, wooded areas, buildings, driveways, and lines of vegetation marking fence lines are usually easy to interpret. Villages or groups of houses, railway lines, highways, and larger commercial or industrial buildings stand out easily. You can probably find your property with very little effort by examining the landscape around you and the photo. The example below shows rural land at the edge of a small village in southern Ontario.

An Example of an Aerial Photo

This aerial photo shows a rural landscape with a 50-acre (20 ha) parcel of woodland and old farm fields in the center. A river flows north on the left side; a small village is located in the northeast corner. In the past this area was used mostly for farming; now young trees are growing in many former fields as natural succession takes over. Try to pick out

- the roads on the south and east sides of the photo,
- the meandering light-colored line of the river,
- the houses of the village, and
- an orchard with regularly spaced trees on the west side of the road, just at the south edge of the village.

On the photo a 50-acre (20 ha) parcel of land is marked. This is a typical property that a rural landowner might manage as a woodland. Although some of the land looks like open fields, the photo was taken in 1978. Today young trees are growing in the old fields. See if you can also spot

- the road on the east side of the property,
- the dark line of a pine windbreak along the road,
- a larger forested area in the west portion of the land, and
- the river along the west edge of the property.

On the aerial photo you can easily pick out the woodland area and fields, but it would be difficult to guess what types of trees were growing in this woodland. To get that information, you need to inspect the area on the ground.

Understanding the Scale of Aerial Photos

It is really useful to understand the scale of aerial photographs and maps. The original of the photo above was taken at a scale of

1:10,000. This means that every inch (or other unit) on the photo is equivalent to 10,000 such units on the ground, or that the photo is one ten-thousandth the size of the real landscape, in terms of measured distances.

If you are working in metric, this makes measurement of distance very easy, because 1 centimeter on the photo is equal to 10,000 centimeters on the ground. This translates to 1 centimeter on the photo equaling 100 meters on the ground (or 1 millimeter equaling 10 meters). You can then calculate distances between any objects you can see on the photo, and check them against the real thing.

An area 1 centimeter by 1 centimeter on the photo represents 1 hectare on the ground. Knowing this, you can calculate the approximate area of your own woodland. A similar scale can be used to calculate distance or areas on any aerial photo, in either imperial or metric units.

This is one point at which an expert might be very helpful; they can explain the scale and interpretation of an aerial photo so you can make use of it in your own woodlot management.

The Rough Sketch Map

Whether or not you use an aerial photo to help you determine the shape of your woodland or the ecological communities within it, you can best express this stage of your inventory by drawing a sketch map. You may not be a professional mapmaker, but it's really very easy. Simply draw the overall shape of your woodland, then divide it into what you have decided are the major ecological communities.

The following example illustrates the detail you might show on a sketch map at this stage. This particular map is for the 50-acre (20 ha) property shown in the middle of the aerial photo on page 51. It shows that more than half the land is forested, with the rest of the property being old fields now regrowing with trees, except for a small orchard the owner has planted in the northeast corner. There are three ecological communities inside the woodland: an area of young maples on a high ridge, older maples on the flood-plain beside the river, and an area of cedar trees that occupy a steep slope. As you gather more details on these areas during your inventory, you can add to this description.

If you compare this to the aerial photo on page 51, you'll see the woodland area and the windbreak along the road, but the orchard and the shack did not exist when the photo was taken. You always

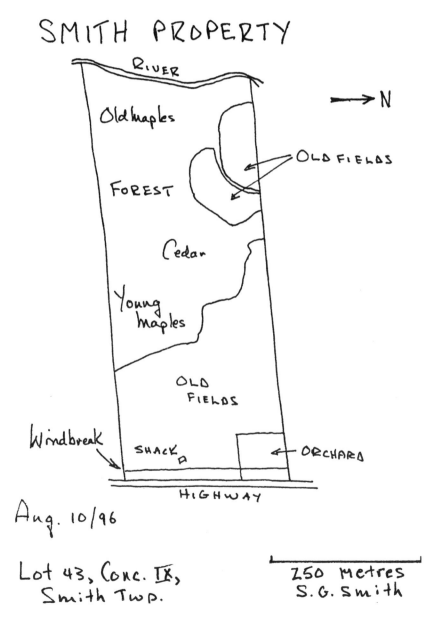

need to combine the use of an aerial photo with real field work. This property is simple enough that you could probably draw the map without the aerial photo, though you might not portray the shapes as accurately. As long as you can recognize the communities when you are on the ground, this does not matter too much. If you cannot easily acquire an aerial photo, do not let that prevent you from developing your own rough sketch map.

To be technically correct, as in cases where you are using a sketch map to qualify for a government support program, your map should always include

- a north arrow,
- the location,
- your name,
- the date, and
- a scale.

Measuring Distances and Finding the Size of Your Woodland

"Pacing" is a handy and simple way to measure approximate distances on your land. A "pace" is simply a double step; for example, counting every time you put your right foot down. You can find out how long your own pace is by measuring out 100 feet (30 m) in your woodlot and pacing it, counting every other step. Dividing the number of paces you take into 100 feet (30 m) will give you the length of your personal pace. Then after counting paces in your woodland, you multiply to get the actual distance.

For example, my pace (2 steps) is 5 feet (1.5 m); if I walk 60 paces, I know that the distance I have covered is approximately (60 x 5) 300 feet (90 m).

For almost all purposes in managing your land (except building fences), pacing is an accurate enough measure for a landowner's needs. If you want to be more accurate, a tape measure or a 50- to 100-foot (15–30 m) light nylon rope marked every 10 feet (3 m) with a permanent black marker is better. For accuracy, measuring with such a rope requires two people.

If you do not know the size of your woodland, it can be figured out in two ways: by walking and pacing or measuring, or by using an aerial photo.

Since an acre is approximately 208 feet by 208 feet, you can pace the approximate distances along the edge of your woodland and calculate how many such squares would fill the woodland. Similarly, you can calculate the area in hectares, knowing that a hectare is approximately 100 meters by 100 meters.

To measure the area of your woodlot using an aerial photo, just measure the two dimensions of the woodland to the nearest 16th of an inch, or to the nearest millimeter, on the photo. Pretend your woodland is a series of small squares or rectangles, if necessary, and add them up after calculating them separately.

Using a 1:10,000 aerial photo as an example, each millimeter on the aerial photo equals 10 meters on the ground, so simply multiply your distances in millimeters on the photo by 10 to get your distances in meters on the ground. Multiply the number of meters on two sides to get the area in square meters. Then divide by 10,000 to translate this into hectares. You can follow the same process using fractions of an inch on the photo, and distances in feet on the ground, to calculate the area in acres.

The following conversions may help:
- 1 acre = 43,560 square feet
- 1 hectare = 10,000 square meters
- 1 acre = .405 hectares
- 1 hectare = 2.47 acres

Step 3: Identify the Trees

To fill in your sketch map you require a little information on what trees and other plants are present in each woodland community. You do this by spending time in each of the ecological communities that you found and identifying the three or four most common trees.

Identifying trees is easily done by using a field guide. There are a few key characteristics you need to learn, but after that, with a little care, anyone should be able to figure out what tree species they are looking at. Key characteristics include

- opposite or alternate leaves and twigs,
- simple or compound leaves,
- single-toothed or double-toothed leaves,
- clustered or single needles, and
- lobed leaves.

opposite *alternate* *simple* *compound*

single-toothed *double-toothed* *lobed*

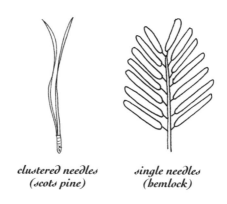

clustered needles
(scots pine)

single needles
(hemlock)

Basic leaf characteristics are easy to recognize. Knowing these and with the help of a field guide, you can learn to identify your woodland trees quickly.

All of these are very easy characteristics to learn. In fact, we believe that you can easily learn to identify all the common trees in your woodland, even in the winter.

It is quite easy to learn to identify most trees without their leaves. For most common species, twigs and bark are quite distinctive. Watching for whether twigs are opposite or are placed alternately on the branch, and learning some bark patterns can help you.

If you spend some time in the early fall when the leaves are still there to help you, then return later to compare the bark of those trees you have identified, you will soon become an expert.

Recording Information

In order to fill in your woodland inventory, you need to record the major tree species in each woodland community. For example, if you did your field work on the property shown on the aerial photo and rough sketch map a few pages back, you would ultimately find the following five communities:

1. Young sugar-maple stand: sugar maple, white ash, some old hawthorn; located on high ground. Many trees are only 2 to 4 inches (5–10 cm) in diameter.
2. Eastern white-cedar grove: composed entirely of a dense cedar stand. Many trees are quite young.
3. Floodplain sugar-maple stand: older sugar maple, some basswood, butternut, hop hornbeam; located on a low floodplain by a river. Trees are from 6 to 12 inches (15–30 cm) in diameter.
4. Regenerating old fields: dominantly white-ash seedlings, with some sugar maple and elm. Planted white pine, red oak, and walnut scattered in patches through fields.
5. Scots-pine windbreak: a narrow strip of forty-year-old Scots pine, now about 6 to 8 inches (15–20 cm) in diameter and 50 feet (15 m) tall, bordering the road.

*Keep some field notes:
You may think you can
remember everything you see
when walking through the
woods, but you won't. Keep
some field notes with the
dates and list of what you see
so you don't waste your
efforts, and you will have the
information you need when
you need it. Take some pic-
tures; collect and press some
leaves if you want to check
your identifications.*

To prepare the basic inventory, you don't need any more detail than this, but you can add as much information as you wish. On this property you would also find two tiny intermittent streams, a low-lying floodplain area along the river, and a fox den. The owner maintains a mown tractor trail down through the old fields to the river and has seen white-tailed deer on numerous occasions. Compare this list to the rough sketch map and aerial photo presented earlier to help yourself decide how you should do this for your own land.

With your information on the trees present in each vegetation community, you can fill this in on your rough sketch map. Take a look at the final sketch map on page 70 to see how this is done.

Step 4: Add Information on Cultural, Physical, and Special Biological Features

Cultural Features

Mark on your sketch map any features that reflect human use of the woodland, such as

- roads, trails, and access points,
- fences,
- buildings,
- old foundations, or
- old garbage dumps.

All these special features help you make decisions about management options after you have finished your inventory. Don't worry about a lot of detail at first; you can always go back later and add more if you wish.

Physical Features

Record the locations of features, such as

- drainage channels, streams, or rivers,
- any wetlands, ponds, or springs, or
- outcrops of sand, rock, or gravel.

Add a few words to describe different topographical features: rolling, flat, hummocky, gently sloping, or steep. Both steep slopes and wet areas are important features of your woodland. They add to diversity and are features that should be considered carefully in relation to management activities, such as timber harvesting or trail building.

It is very useful to put on your raincoat and boots and walk your land during or after a heavy rainfall, and in early spring when the water table is still high. Areas that show up as wet under these conditions should be protected as part of the drainage system, even though they may be very dry in August.

One of the most important physical features of your land is its soil. Add a good description of soil conditions, especially if you are planning to plant trees.

SOIL MOISTURE The most important soil condition is moisture, which is closely related to soil texture. Extremely wet conditions and extremely dry conditions are both a limitation for growing trees, though they may add significantly to the diversity of your land, providing habitats for unique species. Trees, like corn and other crops, grow best in deep loam soils.

SOIL DEPTH The second important characteristic of soil is its depth, especially when the water table or rock is at a fairly shallow depth below the surface.

Take a shovel and examine the soil any place you are planning to plant trees. Look for the kind of soil you expect to find in a garden, and hope that it is easy to dig.

There are four basic soil types to watch for.

1. Sand and gravel: coarse particles that water will run through quickly; therefore, sites with sand and gravel are usually dry, unless located in a low area such as a floodplain.

2. Loam: a mixture of all sizes of particles, with a dark brown color below the surface (top soil). These sites are usually the best for tree growth, just like a vegetable garden.

3. Silt and clay: fine particles that stick together, especially when wet. These sites are usually wet and poorly drained because the water cannot get through the closely packed particles. Hard to dig, especially if dry.

4. Organic soil: organic soil is black muck, composed entirely of decayed vegetation material; it is only present in a wetland area and severely limits the types of trees, if any, that can be grown. Such areas are better left as natural wetlands.

CHECKING SOIL TEXTURE In order to determine what type of soil you have, you can perform your own soil-texture tests by taking a lump of soil in your fist. Use the following criteria to compare soil-texture types:

1. Sand and gravel: a clump of this soil will likely fall apart unless it is very wet; you cannot toss it from hand to hand. You can see individual grains of sand and gravel.

2. Loam: this soil will readily form a clump, but it will fall apart if you try to play catch with it or if it is really dry. Compare it to the soil you'd like to find in your garden.

3. Silt and clay: a clump of this soil will form a solid lump, especially if it is a little wet. You can toss it from hand to hand with no problem; if it is pure clay you could probably play catch with it.

4. Organic soil: you won't want to handle this soil in your hands; it consists of wet, black decayed vegetation. If you do handle it, it will leave your hands black, dirty, and hard to clean.

MOTTLES AND GLEYING Mottles are small specks or patches of color in the soil, usually yellow or orange. They are noticeably different from the rest of the soil in color, though they may be very small, perhaps only a millimeter or two in size. They indicate that the soil is poorly drained at some times of the year.

Gleying is a grayish color of the soil caused by the soil being saturated with water. This can vary from a slight gray color in patches to a distinct gray color through all the soil. Usually this occurs in clay-textured soils. It indicates that the soil is saturated with water for much of the year. These characteristics help you recognize poor drainage, even though the soil may be dry at the time.

LIMITATIONS FOR TREE PLANTING

1. Sand and gravel: creates some limitations for tree planting; only certain species will survive on these coarse dry soils because they do not hold moisture well.

2. Loam: deep loam soils on flat or gently sloping land provide our best quality land for growing trees. They hold soil moisture and are relatively easy to dig.

3. Silt and clay: serious limitations for tree planting; only a few tree species will survive. Planting these wet soils may require special treatment, since water will sit on top of clay soil for some time before infiltrating.

4. Organic soil: organic soils are better left in wetlands to regenerate naturally.

5. Mottling or gleying: can also provide serious limitations for tree growth, especially gleying, because it is associated with wet conditions.

There will likely be a mixture of soil textures on your land. Some clay or some sand and gravel in the soil is not necessarily a serious limitation. Watch for extreme conditions of very dry or very wet. A nearby neighbor who farms the land or has had experience with planting trees may be able to help you assess soil conditions.

Try to include some brief description of soils in your woodland inventory, especially if you plan to plant trees, since soil conditions will determine the growth rate as well as the survival of tree species. You need a reasonably detailed soils description for reforestation planting, showing variations from wet to dry within the area to be planted.

Special Biological Features

You can also list any special biological features you find on your land. List birds, mammals, or other wildlife that you see, as well as habitat features. Add interesting plants such as spring wildflowers or any rare species that you can identify. If you are interested, you may want to buy a field guide to identify wildflowers, birds, or other features.

Habitat features to look for:
- dead trees (snags)
- den or cavity trees
- fallen logs and woody debris
- rock and brush piles
- dens in the ground
- wildlife tracks or travel routes
- supercanopy trees that stand out above the others
- stick nests of hawks or owls
- fencerows
- natural vegetation along streams and around wetlands and ponds
- springs, streams, and wetlands or other surface water
- stands of coniferous trees
- trees and shrubs that bear nuts or fruit ("mast"-producing trees)
- trees that are unique because of their size, species, or growth form

Large old trees providing dens for wildlife are a critical component of wildlife habitat.

Standing dead trees or snags are home to over fifty species of birds and mammals. Smaller birds take over woodpecker nests; mammals live in larger hollow trees. Old white pine are often "supercanopy" trees, standing out above the others. They are important perching sites for hawks and owls. Predators are an important part of natural ecosystems. Isolated areas of land may provide useful sites for the dens of animals such as foxes.

Brush piles, old groundhog holes, and rock piles may provide homes and shelter for animals such as rabbits. You can artificially build such homes, either as protected holes in the ground or as piles of branches. Remember to place some large logs or rocks at the base of a brush pile to provide the needed living space.

You can also include in your inventory any wildlife sightings that you or your family have seen or recorded in the past. Seasonal

Fallen logs provide habitat for numerous insects, birds, reptiles and amphibians in the woods. Without them intermediate levels of the woodland food chain may be missing.

patterns of wildlife records are important. Nests, tracks, and droppings provide other records of wildlife presence. Exploring your land in all seasons will help you learn to identify the wildlife that may be present.

Several species of birds, such as the ovenbird, only nest successfully inside large forest areas. Protecting any particularly large woodland is an important contribution to conserving these species, known as "forest-interior bird species." You may also want to watch for rare or endangered species; there are a number of species of plants, animals, birds, and others that are rare, threatened, or endangered, and deserve extra protection.

If your personal interest is in conserving your woodland as a natural ecosystem, in understanding the woodland ecology, and in nature appreciation, there are other approaches to your woodland inventory that may help you.

1. Spend time in your woods in early spring, especially the first three weeks of May. Take a pair of binoculars as well as a field guide, and watch for spring wildflowers and birds. Try shutting your eyes and listening to the

bird calls. You'll be amazed at the diversity of life in your woods.

The best bird listening is in the early morning, even before sunrise, so take your walk early. Go back several days in a row when new plants begin to emerge and the weather is warm. Watch how they unfold and grow, sometimes inches each day. Once the mosquitoes come out in late May, leave the woods alone for the birds to nest in and successfully raise their young without disturbance.

There are numerous field guides for identifying birds, as well as tapes and videos that will assist you in recognizing bird calls.

2. Listen to the sequence of frog calls in the spring, beginning in early April. About eight or ten species are distinguishable once you know the calls, and the different species are heard in sequence over a period of about two months. We have identified at least seven species in the wetland behind our house in the spring.

State or provincial wildlife agencies can tell you where to acquire a tape of frog and toad calls.

3. Walk your woods a few times in the winter and watch for animal tracks, especially in early and late winter when animals are more active. The best conditions are after a light snowfall on top of a hard crust of snow. A field guide to animal tracks will help you identify what you are seeing.

4. Ask the local naturalists' club or relevant state or provincial agencies to recommend a biologist who will walk through the woodland with you to help you learn more about the plants or birds. Members of a local naturalists' club will often be delighted to share their knowledge with you, as will other knowledgeable landowners.

5. Ask the relevant state or provincial agency if your woodland is one that has been evaluated for biological significance or for rare species. If it has, ask them to provide you with a copy of the results of any field studies. You might want to hire a biologist to undertake a detailed ecological inventory of your woodland, especially to look for any rare species.

You might also consider hiring a third- or fourth-year university student enrolled in an appropriate program; he or she can do a general woodland inventory, and they certainly need the work.

CHECK YOUR WOODLAND FOR INDICATORS OF OLD-GROWTH FOREST Recent research on the characteristics of "old-growth" forest has generated new insight into the features that contribute to healthy ecosystems in the woodlands of eastern North America. Old undisturbed forests in eastern North America tend to have the following characteristics:

- numerous large trees, greater than 20 inches (50 cm) in diameter
- only a few species, especially beech, sugar maple, and hemlock, in the canopy
- significant numbers of large decaying logs (greater than 16 inches [40 cm] in diameter) on the ground
- "pit and mound" topography, the uneven forest floor caused by falling trees and their gradual decay
- plentiful spring flowering plants
- numerous species of mosses on the bark of trees
- significant numbers of large wildlife trees, usually dead snags with cavities or dens
- bird species requiring large forest-interior habitat
- if the habitat is large enough, large carnivores such as black bear, eastern cougar, wolf, or bobcat
- a large size, a minimum of 185 acres (75 ha) for forest-interior birds and much larger for large mammals

Selected Species in the Old-Growth Forest Ecosystem

BIRDS	MAMMALS	AMPHIBIANS AND REPTILES
white-breasted nuthatch	*Southern flying squirrel*	*red-backed salamander*
pileated woodpecker	*raccoon*	*gray tree frog*
downy and hairy	*porcupine*	*spring peepers*
woodpeckers	*fisher*	*spotted salamander*
hermit and wood thrush	*marten*	*wood frog*
ovenbird	*hoary bat*	*red-bellied snake*

PLANTS	
bloodroot	*blue cohosh*
Canada mayflower	*dutchman's breeches*
hepatica	*spring beauty*
trillium	*trout lily*
ferns	*mosses and liverworts*

It will add significantly to your inventory if you are able to note and describe detailed biological features such as these. The discovery that your woodland includes many of these features may be an important influence on your consideration of management options.

RARE, THREATENED, AND ENDANGERED SPECIES The recognition of rare species of plants, animals, or other life forms requires specialized knowledge. Scientific-research programs include studies documenting these species, and many nongovernment groups work to encourage their protection.

Before we can say that an individual species is rare, it's distribution must be known. One of the most interesting volunteer environmental projects in recent years has been the work done to create atlases of birds, mammals, reptiles, amphibians, and other species. The knowledge gained in these broad studies enables the identification of individual species that are found in only a few localities. Scientists use all these records to determine whether to designate a species as rare, threatened, or endangered.

A sampling of these species in eastern North America includes the following:

ENDANGERED SPECIES	RARE SPECIES	THREATENED SPECIES
eastern cougar	southern flying squirrel	spiny soft-shell turtle
peregrine falcon	barn owl	American ginseng
blue racer	red-headed woodpecker	Kentucky coffee tree
cucumber tree	red-shouldered hawk	
eastern prickly pear cactus	short-eared owl	
small white lady's slipper	wood turtle	
small whorled pogonia	spotted turtle	
	green dragon	
	calypso orchid	
	linear-leaved sundew	
	mountain heather	
	ram's head lady's slipper	

The species considered rare in any one region will vary across eastern North America. Tulip trees are rare in Canada because they are at the northern limit of their range, but they are common in Tennessee. Detailed biological records are usually kept by state or provincial agencies or nongovernment organizations such as the Nature Conservancy (U.S.) or the Nature Conservancy of Canada. Many states and provinces have established natural heritage information centers where local information on rare species is available.

Step 5: Draw the Final Sketch Map and Summarize Information

To summarize your basic woodland inventory, you should fill in your original rough sketch map showing the ecological communities with more detailed information. Add the cultural features, physical features, and any special biological features. It is a good idea to write out a description of each ecological community on a separate piece of paper, with a list of species you have seen there. Include a description of each, with notes of soils, drainage, and slopes.

The sketch map below is a good example of what you might end up with. Keep it simple and legible. This is the foundation for considering your woodland-management options, so it is well worth doing this thoroughly.

Final Sketch Map

Many landowners like to prepare a more thorough report with some history of their property, photographs, and other information. Sometimes a conservation agency will prepare such a report for landowners. A government support program may require a more detailed inventory report to go along with your sketch map.

For a more detailed report, you can consider including the following:

1. Your name and property location, and possibly a location map
2. The name of the consultant who prepared your inventory or plan, if you did not
3. A general description of the property and its history
4. A description of the surrounding landscape and watershed
5. A general description of the ecological communities you have found and the physical, cultural, and special biological features of the entire property
6. A detailed description of each of the ecological communities you marked on your sketch map, including physical characteristics, history, tree species, other plants and biological features, and wildlife sighted or wildlife-habitat features
7. Detailed inventory information for timber and firewood management, as described in the following chapter

Although the task of preparing a woodlot inventory may seem overwhelming at first, be assured that it is not. Many landowners we have worked with have successfully prepared their own

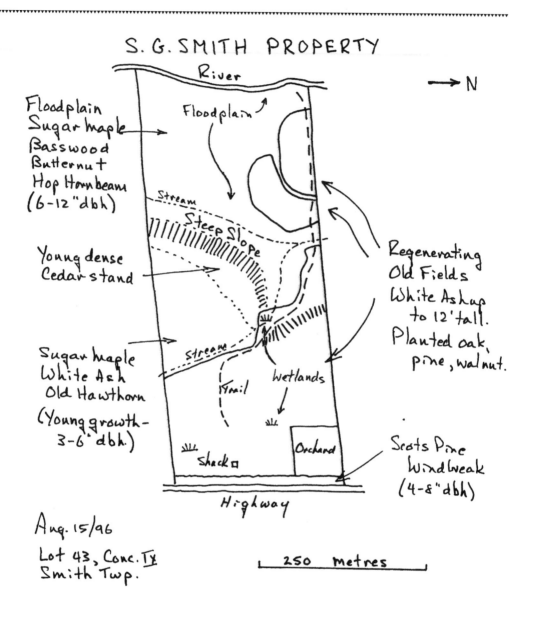

S. G. SMITH PROPERTY

River

→ N

Floodplain
Sugar maple
Basswood
Butternut
Hop Hornbeam
(6–12" dbh)

Floodplain

Stream

Steep Slope

Young dense
Cedar stand

Regenerating
Old Fields
White Ash up
to 12' tall.
Planted oak,
pine, walnut.

Sugar maple
White Ash
Old Hawthorn
(Young growth –
3–6" dbh)

stream

Trail

Wetlands

Shack □

Orchard

Scots Pine
Windbreak
(4–8" dbh)

Highway

Aug. 15/96

Lot 43, Conc. IX
Smith Twp.

250   metres

inventories in relatively little time. It is easier than you think and opens the door to a whole new world of understanding.

In the following chapter, we outline the detailed information that must be gathered to extend this basic inventory to enable management decisions on firewood and timber harvesting. In the next five chapters, we outline a complete range of management options you can consider.

In Chapter 10, we further discuss the preparation of a steward-ship plan based on your inventory, after you have carefully considered your own objectives and all possible management options. This information can then be added to your inventory to complete your woodland-stewardship plan. This basic descriptive inventory is the starting point for all woodlot management.

Further Reading

Beattie, Mollie, Charles Thompson, and Lynn Levine. *Working with Your Woodland: A Landowner's Guide.* Hanover, NH: University of New England Press, 1993.

Decker, D.J., et al. *Wildlife and Timber from Private Lands: A Landowner's Guide to Planning.* Ithaca, NY: Cornell University Extension Service, 1983.

Fazio, J.R. *The Woodland Steward.* Moscow, ID: The Woodland Press, 1985.

Hilts, S.G., and P. Mitchell. *Taking Stock: Preparing an Inventory of Your Woodland.* Guelph, ON: University of Guelph, 1997.

Landowner Resource Center. "Do You Have a Healthy Woodlot" (fact-sheet). Manotick, ON: Landowner Resource Center, 1997.

CHAPTER 4

An Inventory for Firewood and Timber Production

The woodlot inventory described in Chapter 3 should provide a basic foundation for many management decisions. But in order to manage a woodlot for firewood and timber, it is necessary to go beyond this and gather detailed information on the numbers, species, and sizes of trees present, to prepare a timber inventory. You have to actually count, identify, and measure the size of your trees.

If the goal of your management plan is to provide primarily recreation, wildlife appreciation, or cultivation of a young forest, there is no immediate need to do a timber inventory. However, it may be very useful if you want to manage your forest to enhance its old-growth characteristics or to maintain present recreational characteristics. Some government support programs may require that you proceed far enough in your inventory that you calculate "species composition" (described below), even if you never intend to harvest timber.

If you intend to only cut a little firewood for your own use, you will be able to plan this better, with less impact on the health of your woodland, if you base your decisions on a detailed inventory. Specific management techniques can be used to establish and highlight many different forest characteristics and values, whatever those may be. Active management is particularly important if your forest is entirely second growth, if it has been grazed by cattle in the past, or if it has been logged excessively.

Forest Types

There are many different forest types, or associations of trees, that are found in the woodlands of eastern North America. Which major categories are located in your woodland depend on whether it is in an upland (dry) or lowland (wet) location, and whether the forest type is deciduous, coniferous, or mixed. To these criteria can be added deciduous and coniferous plantations, resulting in eight categories of forest types, as follows:

- upland deciduous forest
- upland mixed forest
- upland coniferous forest
- lowland deciduous forest
- lowland mixed forest
- lowland coniferous forest
- coniferous plantations
- deciduous plantations

Within these groups there are other forest types that can be divided on the basis of species. For example, an upland deciduous forest could be dominated by sugar maple or it could be a mix of other species, such as oak, hickory, walnut, or ash.

Among these forest types, the upland stands that support the growth of sugar maple, beech, oak, ash, black cherry, hickory, and other hardwoods are the most valuable for timber. Most of our experience in timber management in this region of eastern North America is with this forest type – the deciduous hardwoods – and it is these trees that are valuable enough to merit the cost of active management. In some cases, conifer stands can also provide valuable timber resources, but the dominant timber harvests of conifers are outside the eastern deciduous forest region.

Numerous other products besides timber and firewood may be gained from other forest types, as noted in Chapter 6, but the basic timber inventory described here is focused on the deciduous hardwood forest. Timber management is discussed in detail in Chapter 6, and management of plantations is discussed in Chapter 7 on reforestation.

Within the deciduous hardwood forest, some trees are more valuable than others. The chart below illustrates the relative values of these species.

Table 2   Tree species by value

HIGH VALUE	MEDIUM VALUE	LOW VALUE
Black walnut	Basswood	Aspen
Black cherry	Tulip poplar	Beech
Red oak	Yellow birch	Hop hornbeam
White oak	White birch	Hemlock
White ash	Red maple	White cedar
Sugar maple	White pine	Butternut
	Red pine	Hickory
	White spruce	Elm

Source: S.N. Staley. *Wood: Take a Stand and Make it Better.* Toronto: Ministry of Natural Resources, 1991.
G.R. Goff, J.P. Lassoie, and K.M. Layer. *Timber Management for Small Woodlands.* Ithaca, NY: Cornell Cooperative Extension Service. 1994.

Soil types and drainage conditions are major influences on the quality of trees that may be present, and therefore their value. On thin, rocky soil, trees may never reach high quality, even if they do include the species listed here.

Planning Your Timber Inventory

Timber inventories are usually undertaken by professional foresters. These experts have the training and experience to do the work quickly and efficiently and charge relatively low fees in comparison with the value of a timber sale. They are often hired by landowners, not merely to do a timber inventory, but to mark the trees for harvest, and to oversee the timber sale and harvesting process afterwards.

However, if you wish to do your own inventory, there are relatively simple techniques you can use.

Three common calculations are made in order to plan management decisions.

1. Species composition
2. A count of the number of trees per acre
3. A calculation of "basal area"

The species composition is the number of important tree species found in a woodland, expressed in percentages. For example, a sugar-maple stand might be 80 percent sugar maple, 10 percent beech, and 10 percent black cherry. You calculate this by counting the trees of each species in a given area.

The number of trees per acre is simply a count of trees expressed in relation to area, a measure that helps you calculate "stocking." The concept of stocking is the idea that if your goal is timber production, a woodlot can be overcrowded or under-crowded. In conducting a timber inventory, you will gather this information in relation to different diameter categories, as described below.

The basal area of a tree is the surface area that the stump would have if you cut the tree off at 4.5 feet (1.5 m) above the ground, as if you were looking down from above on the circle formed by the trunk. The basal area of a stand of trees is the total basal areas of all the individual trees. Measuring the basal area is important because a few larger trees will produce more timber volume than many smaller ones.

Timber volume is a fourth measurement that is eventually calculated during the marking of a woodlot for harvest, or after harvest. It can be roughly estimated in advance to assist in management choices. Volume is the amount of wood in a stand of trees, usually measured in "board feet" – one board foot being a board

1 foot (30 cm) long by 1 foot (30 cm) wide by 1 inch (2.5 cm) thick. Volume is now also measured in cubic meters. For harvesting purposes you are only interested in the volume to be harvested, not the total volume of your woodlot.

Simplified Steps for Landowners

These steps assume that you have already identified the vegetation communities in your woodland; the process must be repeated separately for each vegetation community.

This approach will provide a working estimate. You should consult a professional forester, to at least check your calculations before basing major management decisions on the information you gather yourself.

STEP 1: SAMPLING You can gather the information you need to know by "sampling" small areas within your woodland; you don't have to measure every tree. "Sampling" means measuring specific smaller patches, perhaps 2.5 percent or 5 percent of the total area, then basing your total estimate on this. Try to pick random points to sample rather than unique or unusual spots, or as described below, follow a compass bearing to lay out a regular grid of points.

The easiest sample to measure by yourself is probably a circular plot with a radius of 37.2 feet (11.3 m); this is equal to ⅒th of an acre. If your woodland is 10 acres, five plots would give you a sample ½ acre, or 5 percent. For general management, try for a 2.5-percent sample; but for significant timber harvesting, try for more detail. If you are working in the metric system, use a radius of 11.3 meters for the circular plot; this provides a sample ½₅th of a hectare.

A circular sample is easy to work with in a woodlot. Choose a point and put a stake at the center. Then walk in and out with a rope 37.2 feet (11.3 m) long to count, identify, and measure every tree in the circle. For timber inventory, only count trees larger than 4 inches (10 cm) in diameter. With practice, it's easy to keep track of trees. If you get confused, use a piece of chalk to check off each one as you count it.

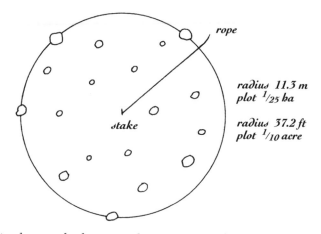

A circular sample plot is easy for one person to lay out and measure with just a measured rope for reference.

Alternatively, you can use square sample plots. A square 66 feet on each side is ⅒th of an acre; a square 20 meters on each side is ½₅th of a hectare.

Using a compass to lay out samples:

To prevent sampling an area twice, and to be sure you cover your entire woodlot, it may be useful to follow a compass bearing when picking sample points.

For example, if your woodlot is oriented from west to east, follow a bearing west (270 degrees), stopping every few hundred feet for a sample. Then move over (north or south) and return on a bearing of east (90 degrees), gathering more samples.

STEP 2: IDENTIFICATION AND MEASUREMENT To complete your inventory, you need to identify the species and measure the diameter of each tree inside your sample area. You should also estimate the height of the straight portion of the tree trunk.

There are several ways to measure the diameter. You can measure the circumference with a regular measuring tape, and divide by pi (3.14), or use the table below to get the diameter. You can also buy special diameter measuring tapes; although you wrap them around the circumference of the tree, the readings give you the diameter. Foresters also use large calipers to quickly make these measurements.

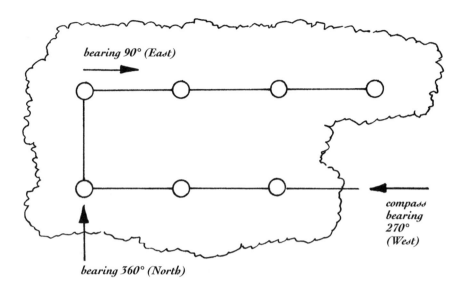

Following a compass route through your woodland can help you accurately sample the vegetation.

Table 3 Diameter table

CIRCUMFERENCE IN INCHES	DIAMETER IN INCHES	CIRCUMFERENCE IN CENTIMETERS	DIAMETER IN CENTIMETERS
12.5	4	31.5	10
19	6	44	14
25	8	56.5	18
31.5	10	69	22
38	12	81.5	26
44	14	94	30
50.5	16	107	34
56.5	18	119	38
63	20	131	42
69	22	144.5	46
75.5	24	157	50

To calculate exact results, divide the measured circumference by the standard value 3.14 or 'pi'.

Perhaps the easiest measuring technique is to make your own cruising stick. This is simply a yardstick that you can hold up against the tree trunk to estimate diameter; however, like a diameter tape, it is calibrated with measurements of diameter rather than true distance. On the other side it can be calibrated to measure height. The instrument with calibrations used to estimate diameter is referred to as a "Biltmore stick"; the instrument used to estimate height is referred to as a "Merri hypsometer."

To use a Biltmore stick, hold it up horizontally against the tree trunk 4.5 feet (1.4 m) above the ground. Be sure to hold your shoulders square and the stick horizontal. Move the stick until your eye lines up the left end of the stick with the left side of the tree trunk; then without moving your eye, read off the diameter where your eye lines up with the right side of the tree trunk. The correct markings for a Biltmore stick and hypsometer are provided in Table 4 on page 83.

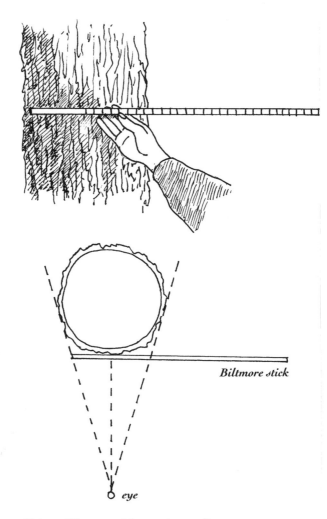

Biltmore stick

eye

Using a Biltmore stick to estimate diameter.

Whatever method you choose, measure the diameter of every tree within your sample plot and identify the species. A chart for recording this information is provided on page 86.

Professional foresters also use a "prism" to quickly estimate the total basal area at a sample point. This avoids having to measure out a sample plot at all.

A "prism" is a small piece of glass ground to a pre-determined angle. When you look through it, tree trunks are slightly offset. To do a prism sweep turn in a 360-degree circle keeping the prism above a set point on the ground (turn your body, not the prism). Trees that are offset by more than the width of the trunk are not counted, but all trees where the trunk as seen through the prism is offset by less than the trunk diameter are counted. You must also record the species and diameters of trees that you count. Having a second person to do this as you record the information is extremely useful.

Each prism will have a set "basal-area factor" to calculate basal area in either square feet or square meters. Simply multiply your count of trees by this factor to establish an estimate of basal area per acre or hectare. For example, if you find fourteen trees in your sweep with a 2X-metric prism, multiply 14 by 2 to get 28. Because your prism is in metric units, this indicates 28 cubic meters per hectare.

If you are managing a commercial-scale woodlot, a prism is a worthwhile investment, since it speeds up calculations considerably.

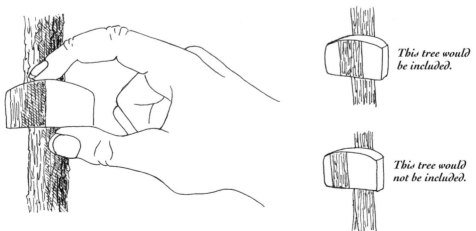

This tree would be included.

This tree would not be included.

Using a prism to estimate basal area.

The biggest benefit is that you do not have to lay out a sample plot at all; the trees that are counted during the prism sweep are your sample.

Whatever technique you use, don't guess; unmeasured guesses vary widely from reality. Professionals measure every time; but for timber management purposes, measurements have to only be accurate to about the nearest inch (2.5 cm).

You may also want to measure tree heights or the height of potential sawlogs. This is not needed to calculate stocking of your woodland, but it may help you estimate eventual timber volume.

To measure height using a Merrit hypsometer, you must stand a set distance, such as 66 feet (15 m) away from the tree. Hold the stick vertically, and line up the zero reading with the base of the tree. Then read off the height of the tree to the first major branches. A hypsometer commonly has the zero reading about 6 inches (15 cm) above the end of the stick, so you can hold it with your fist below this point. The reading is commonly given as the number of 8- or 16-foot (2.5 m or 5 m) logs that can be cut from the tree, rather than in actual feet or meters.

If you are measuring the sawlog to prepare for a timber harvest, measure the straight trunk until the first major branches; if you are

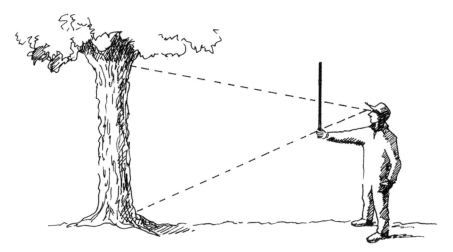

Using a hypsometer to estimate height. You must stand a pre-determined distance from the tree – see Table 5.

interested in the total height of the tree, you need to measure to the top of the canopy. You do not really require height measurements in order to make basic management decisions; this information can be gathered if and when the woodlot is marked for timber harvesting.

STEP 3: NOTING TREE QUALITY As well as measuring tree diameters and heights and identifying species, you should rate the larger trees for their quality. The basic choice here is between trees that are good enough quality to be sawlogs and those that are not. A tree will not be acceptable as a sawlog if it has

- a crooked stem,
- fungus growth,
- holes or cavities,
- large branches low on the trunk,
- dieback in the crown, or
- cracks or interior rot.

Table 4 Biltmore stick markings

DIAMETER TO BE REPRESENTED		DISTANCE TO MARK ON BILTMORE STICK	
INCHES	CENTIMETERS	INCHES	CENTIMETERS
6	15	5.3	12.7
8	20	6.8	16.6
10	25	8.2	20.4
12	30	9.6	23.9
14	35	10.8	27.2
16	40	12.1	30.4
18	45	13.2	33.5
20	50	14.3	36.4
22	55	15.4	39.1
24	60	16.4	41.6
26	65	17.4	44.1

The left end of the stick held horizontally is the zero point. Put markings on the stick at the indicated distances to represent the diameters. If you are left-handed, you can mark the stick in reverse. The scale will vary slightly with the reach of your arm; check these against actual trees to adjust for your own reach.

Sources: H.V. Wiant. *How to Estimate the Value of Timber in Your Woodlot.* Morgantown, WV: West Virginia University Agricultural and Forestry Experiment Station, 1989.

Landowner Resource Centre. *A True Picture - Taking Inventory of Your Woodlot.* Manotick, ON: Landowner Resource Centre, 1998.

Table 5 Hypsometer stick markings

INTERVAL TO MARK ON STICK TO REPRESENT EACH 8 FOOT OR 2.5 METRE LOG	
IF YOUR ARM REACH IS:	INTERVAL TO REPRESENT 8 FEET
21 inches	2.6 inches
23 inches	2.8 inches
25 inches	3.1 inches
For these measurements, stand 66 feet from the tree.	
IF YOUR ARM REACH IS:	INTERVAL TO REPRESENT 2.5 M.
56 cm.	9.3 cm.
60 cm.	10 cm.
64 cm.	10.7 cm.
For these measurements, stand 15 metres from the tree.	

Mark a zero point about 6 inches from the bottom of the stick, held vertically.
Then mark intervals representing each 8 foot (or 2.5 metre) length according to the chart.
Sources: H.V. Wiant. *How to Estimate the Value of Timber in Your Woodlot*. Morgantown, WV: West Virginia University Agricultural and Forestry Experiment Station, 1989.
Landowner Resource Centre. *A True Picture-Taking Inventory of Your Woodlot*. Manotick, ON: Landowner Resource Centre, 1998.

Poor-quality trees should not be counted in your calculation of timber volume, but may still be useful for

- wildlife-den trees or snags,
- firewood, or
- stocking density.

At the other extreme, very high-quality trees with large, straight trunks may be even more valuable if they are acceptable as veneer logs, one of the most valuable woodlot products. Foresters classify trees as "acceptable growing stock" (AGS) or "unacceptable growing stock" (UGS), depending on their quality.

You'll need the following tools for your timber cruise:

1. a tape measure or Biltmore and hypsometer stick
2. a clipboard, paper, and sharp pencil or pen
3. a tree-identification guide
4. a thin rope 37.2 feet (11.3 m) long and a stake
5. possibly a compass

You can also purchase such specialized tools as tree calipers, diameter tapes, a Biltmore stick, Merrit hypsometer, clinometers, and prisms to assist you with your timber inventory. A local professional can direct you to commercial suppliers.

STEP 4: TALLYING YOUR INFORMATION You can gather your information as a list of trees identified with the required information on species, diameter, height, and quality of each one. You can easily make up a table as illustrated here on which to record this information.

Table 6   Sample timber inventory record

TREE SPECIES	DIAMETER	HEIGHT	QUALITY
Sugar maple	*16*	*32*	*ags*
White ash	*15*	*24*	*ags*
Sugar maple	*24*	*32*	*ugs*
Sugar maple	*15*	*24*	*ags*
Beech	*32*	*16*	*ugs*
Sugar maple	*21*	*40*	*ags*
Black cherry	*15*	*24*	*ags*
Sugar maple	*14*	*24*	*ags*
Hemlock	*18*	*32*	*ags*
————			
etc.			

* 'Ugs' represents unacceptable growing stock; 'ags' represents acceptable growing stock. Measurements in this example are in inches.

If you are doing a lot of sampling, it may be easier to record the information about each tree you count on a tally sheet. This sheet records all information quickly on one page, but the categories may take a little getting used to. Take along a clipboard and paper, and just list all the trees you count, recording the species, the diameter, and the height of each one, as well as notes on tree quality.

Once you have gathered this information, you should probably call in an expert to check and interpret it for you. A professional can help you with the following calculations, as well as explaining what they mean for your management options. If you wish, you can follow

Table 7   Tally sheet

DIAMETER IN INCHES	TREE SPECIES													
4														
6														
8														
10														
12														
14														
16														
18														
20														
24														
26														
28														
30														
32														
––														
––														
etc.														

This tally sheet has room to write seven species in across the top; use different symbols to indicate acceptable and unacceptable quality of trees. At the end, simply add the numbers to get the totals in your sample.

these steps to do the calculations yourself, but we strongly advise that you then call on a professional to help you interpret them.

Preparing Your Calculations

After gathering the information from your sample plot, you should end up with
- the number of trees of each species,
- the diameter of each one,
- an indication of quality, and
- the heights, measured either as actual height or in terms of potential 16-foot (5 m) sawlogs (optional).

This is the information you require in order to assess the timber in your woodlot.

SPECIES COMPOSITION Once you have counted a number of trees and their species, translate these into percentages in order to describe the "species composition." Percentages are fractions translated into a proportion out of 100. For example, if you count fifty trees, and find forty-one sugar maple, six beech, and three white ash, multiply by two and you will get the following percentages:
- 82 percent sugar maple
- 12 percent beech
- 6 percent white ash

The standard formula is to divide the number of trees of a single species by the total number of trees in your sample, then multiply by 100. For example, forty-one sugar maple divided by fifty, multiplied by 100, equals 82 percent.

TREES PER ACRE Calculating the average number of trees per acre (or hectare) requires one more step: you need to know how large the area you sampled was. For example, if you have used five samples which are each $\frac{1}{10}$th of an acre as described above, your total sample when added together will be $\frac{5}{10}$ or $\frac{1}{2}$ an acre. Therefore, multiply your total by two to get the trees per acre.

Usually, the number of trees per acre is recorded in different diameter groups. This information helps in determining whether or not to thin a woodlot; often only trees larger than 4 inches (10 cm) are counted.

Diameter groups are commonly

- polewood of from 4 to 10 inches (10–24 cm) in diameter,
- small sawlogs of 10 to 15 inches (26–36 cm) in diameter,
- medium sawlogs of 16 to 19 inches (38–48 cm) in diameter, and
- large sawlogs of 20 inches (50 cm) in diameter or greater.

You will end up with a simple count of trees in these different classes; for example:

- Polewood 48 trees/acre
- Small sawlogs 20 trees/acre
- Medium sawlogs 23 trees/acre
- Large sawlogs 12 trees/acre

Your count of trees in these different diameter classes can eventually be compared to a standard to establish whether your woodlot could be harvested or would benefit from thinning (see Chapter 6). When compared to such a standard, this example has an excess of larger trees and a deficit of younger trees to provide future growth, so it would benefit from a timber harvest. See the discussion in Chapter 6 for further explanation.

BASAL AREA PER ACRE "Basal area" refers to the cross-sectional area of the tree trunk as if a stump had been cut and was viewed from above. For an individual tree, the basal area is therefore the area of a circle. It is used as a measure because a few large trees may be more important than many small trees. Numbers alone do not count; size is also important.

You can calculate the basal area of each tree using the table below. Then add the basal areas of all trees in a species to get the basal area for each species, and add all these totals to get the total basal area of the stand. Assuming your sample was ⅒th of an acre, then multiply by ten to get the basal area per acre.

Table 8 Basal area table

DIAMETER IN INCHES	BASAL AREA IN SQ. FEET/ACRE	DIAMETER IN CENTIMETERS	BASAL AREA IN SQ. M./HA.
4	.087	10	.008
6	.196	14	.015
8	.349	18	.025
10	.545	22	.038
12	.785	26	.053
14	1.068	30	.071
16	1.395	34	.091
18	1.767	38	.113
20	2.180	42	.139
22	2.638	46	.166
24	3.139	50	.196

Calculate exact basal area in square feet per acre using the formula 'diameter squared x .00545'.
Calculate exact basal area in square metres per hectare using the formula: 'diameter x .00007854'.

Management decisions are usually based on the total basal area, but for timber sales, buyers may also want to know the proportion of each species.

You can also calculate basal area by adding up the number of trees in each diameter class and multiplying by the basal area per tree from the chart above. The following chart provides an example of this:

Table 9 Sample basal area calculations

TREE DIAMETER (INCHES)	NUMBER OF TREES	BASAL AREA PER TREE (SQ. FT.)	TOTAL BASAL AREA (SQ. FT.)
5	2	0.136	0.272
6	5	0.196	0.98
8	2	0.349	0.696
15	2	1.226	1.226
18	3	1.767	5.301
24	1	3.139	3.139
Total	15		11.61

If this sample plot represents one-tenth of an acre, the basal area per acre is 116.1 square feet, an overstocked stand worth considering for thinning.

The figures you derive for basal area per acre (or hectare), along with the figures for the number of trees in different diameter classes, are the most important figures you need to make management decisions. The basal-area figure will tell you the stocking level of your woods; management choices based on this are discussed in Chapter 6.

THE PRISM SHORTCUT If you use a prism to measure basal area in your woodland, you can avoid most of these calculations, because it provides a direct basal-area measurement, which can then be compared to a standard (see Chapter 6) in order to make management decisions. However, a prism estimate alone, with no measurement of tree diameters, is a rough working estimate only. It does not tell you the proportion of trees in different diameter classes, unless you also gather this information as you go. In that case, you will need to do some of the above calculations anyway.

VOLUME PER ACRE The volume per acre (or hectare) is usually calculated during the marking of a woodlot for harvesting; this is discussed in Chapter 6. You can make a volume estimate in advance by following the method outlined there, but remember that you will never harvest the entire timber volume in your woodlot, and you cannot decide on what proportion to harvest until you have finished your calculations and compared these to standard prescriptions.

If all this work sounds like too much, you can easily hire a professional to do your woodlot inventory for you. Understanding the process and terminology will assist you in working with them, even if you do not do the field work yourself.

An easy way to start is to simply try one sample plot in your woods. It may not be totally representative, but you can work out all the measurements described above using one sample; this will help you learn and understand the process, which you can then apply more widely. Don't base management decisions on just one sample point.

An Inventory for Firewood Harvest

Firewood is properly harvested in two ways. If firewood is being cut with no associated timber harvest, select the trees on the basis of a thinning plan or cut trees that are unacceptable growing stock for future timber. If a timber harvest is being undertaken, firewood is produced from the unused treetops.

While timber is usually measured in board feet, firewood is measured in "cords" or "face cords." A standard cord (sometimes called a "bush cord") is a pile of wood 4 feet (1.2 m) high, by 4 feet (1.2 m) wide, by 8 feet (2.5 m) long. A face cord is a pile 4 feet (1.2 m) high by 8 feet (2.5 m) long, but only 16 inches (40 cm) deep. Firewood dealers will sometimes sell even smaller cords of pieces only 12 inches (30 cm) long.

The number of cords to be harvested in your woodlot can be calculated in two ways.

If you plan to cut trees for firewood only, use the table below to estimate firewood volume. Even if you are not harvesting sawlogs, it makes sense to plan your firewood harvest to optimize sawlog growth in case you ever want to harvest them. This means that the trees you cut for firewood should be chosen on the basis of a thinning plan. This will do minimum damage to your woodland, while maximizing both old growth and sawlog production. The development of a thinning plan is explained in Chapter 6; it is based on the same calculations described above for planning a timber harvest.

Table 10   Firewood volume table

TREES REQUIRED TO YIELD ONE STANDARD CORD		
DIAMETER	DECIDUOUS TREES	CONIFEROUS TREES
5	35	—
6	20	—
7	15	20
8	11	13
9	8	10
10*	6	8

* Trees this large or larger should be evaluated as potential sawlogs and should only be cut for firewood if deformed or damaged. All measurements are in inches.

Sources: S.N. Staley. Wood: *Take a Stand and Make It Better*. Toronto: Ministry of Natural Resources, 1991.

G.R. Goff, J.P. Lassoie, and K.M. Layer. *Timber Management for Small Woodlands*. Ithaca, NY: Cornell Cooperative Extension Service. 1994.

If you expect to cut firewood from the tops of trees harvested as sawlogs, count on 1.5 cords for every 1,000 board feet of timber in the logs – this would take about three 19-inch (48 cm) diameter trees.

Heat Value by Species

When cutting firewood, it also makes sense to choose species according to their relative heat values.

Table 11   Heat values of tree species

HIGH	MEDIUM	LOW
Hickory	White birch	Aspen
Oak	Red maple	Basswood
Ash	Elm	Willow
Sugar maple	Tamarack	Pine
Beech		Spruce
Hop hornbeam		
Apple		
Yellow birch		

Source: D. Van Ryn and J.P. Lassoie. *Managing Small Woodlands for Firewood*. Ithaca, NY: Cornell University Extension Service, 1987.

Your choice of species should also reflect their value as sawlogs, so compare this chart to that of tree values for timber on page 75. You will see, for example, that beech and hop hornbeam (also widely known as ironwood) do not rank highly for their timber values, but provide high heat values. Sugar maple, white ash, and oak are more valuable as sawlogs, unless they are defective in quality. You should never cut valuable trees such as black walnut or black cherry just for firewood.

A Seasonal Timetable

Your woodland is constantly changing over the seasons. When preparing a description of your woodland, it is important to visit it in different seasons. A suggested timetable for preparing a woodland inventory and management plan follows. Remember, there is no rush to get out the chain saw. Take your time and gather the information you need and seek advice from professionals before making any management decision. Do not let your judgment be rushed by an enthusiastic timber buyer.

Winter, year 1: contact government agencies to obtain any available information on your woodland, including soils, forest types, and topography. Purchase an aerial photo; determine local, state, or provincial regulations governing woodland management.

Early spring: walk through your woods to see differences in physical conditions, especially drainage, and start identifying ecological communities. Determine boundaries if in any doubt.

Mid spring: start identifying trees, birds, and wildflowers in each community; prepare your rough sketch map of communities.

Late spring and summer: leave woods alone for nesting birds (and bugs).

Early fall: finish identifying vegetation while leaves are still on the trees. Invite professionals to walk your woods with you if you have questions you need help with.

Mid to late fall: complete timber inventory with diameter and height measurements when you can easily see through the woodland. Finalize your sketch map. Calculate the size of your woodland, if needed.

Late fall and winter: write your inventory and consider management options. Share your inventory with an expert to evaluate those options.

Spring, year 2: undertake any planned tree planting, nest-box building, etc. Add to your list of spring wildflowers and birds; learn the calls of mating frogs and toads. Hire a professional to mark your trees for any scheduled firewood or timber harvest. Advertise timber sale and establish a written contract with the logger. Get at least two prices.

Late spring and summer: leave woods alone for nesting birds (and bugs).

Late summer, fall, or winter: when ground is frozen or dry, undertake timber or firewood harvesting.

This chapter has been written particularly for landowners interested in timber harvesting, but there are many other reasons for undertaking some cutting in a woodlot. If your interest is in restoring a natural ecosystem or in growing old, large trees, some active management can greatly hasten this process. If your woodlot has been degraded by grazing, species composition can be altered through harvesting timber or firewood. If you need to promote the regeneration of young trees, timber harvesting will help with this as well. The following chapters will help you understand your management options.

Further Reading

Fazio, James R. *The Woodland Steward*. Moscow, ID: The Woodland Press, 1985.

Goff, Gary R., James P. Lassoie, and Katherine M. Layer. *Timber Management for Small Woodlands*. Ithaca, NY: Cornell University Extension Service, 1994.

McEvoy, T.J. *Introduction to Forest Ecology and Silviculture*. Burlington, VT: University of Vermont, 1995.

Staley, R.N. Wood: *Take a Stand and Make It Better*. Toronto: Queen's Printer, 1990.

Van Ryn, Debbie M., and James P. Lassoie. *Managing Small Woodlands for Firewood*. Ithaca, NY: Cornell University Extension Service, 1987.

CHAPTER 5

Environmental Sustainability and Habitat Conservation

Most of the landowners we have spoken to value their woodlands primarily as a place of natural beauty and a habitat for wildlife. Among those owners who manage their woods for timber or firewood harvest, much of the enjoyment of being a woodland owner also comes from spending time in the natural environment, seeing wildlife, and appreciating the woodland for its own sake.

There are many specific techniques landowners should be aware of to ensure that their woods are maintained in an environmentally sustainable manner.

Some of these are obvious, such as preventing soil erosion and conserving wetlands. Others are conventional wildlife management techniques that have been known for many years, and discovered or developed by professional wildlife managers. Most of these techniques were originally developed in order to encourage habitats for game species such as white-tailed deer, cottontail rabbits, or ruffed grouse. However, science is advancing rapidly, and a much more complete picture of wildlife conservation is now being developed, captured by the term conservation of biodiversity.

On one hand, leaving the woods alone is usually an appropriate management strategy for conserving biodiversity and wildlife habitat. Many landowners will be completely satisfied with this. We list practical suggestions in this chapter, not to suggest that you must do something, but to provide some choices if you wish to. We believe that some active involvement in your woodland increases your personal understanding and commitment to conservation.

On the other hand, protecting habitats does not mean that you have to leave your woodland alone year-round. Birds, mammals, and amphibians are most susceptible to disturbance during the spring and early summer months when they are establishing nests, laying eggs, or raising young. Fortunately, this coincides with the time mosquitoes and blackflies are at their worst, so staying out of the woods during the breeding season is a good idea from both points of view. Drainage is most easily disrupted during this time period as well. If done with care and understanding, timber and firewood harvesting can usually be combined with conserving biodiversity.

Among the environmental challenges facing humankind, the protection of habitats for the remaining species on earth is a high priority. As one eminent biologist has written,

> The worst thing that will probably happen – in fact is already well underway – is not energy depletion, economic collapse, conventional war, or even the expansion of totalitarian governments. As terrible as these catastrophes would be for us, they can be repaired within a few generations. The one process now ongoing that will take millions of years to correct is the loss of genetic and species diversity by the destruction of natural habitats. This is the

folly [for which] our descendants are least likely
to forgive us.
— E.O. Wilson, *Biophilia*, 1984.

As this quote highlights, biodiversity conservation is a concern
for the future of all species on earth. Many would argue that other
species have a right to exist apart from any benefit they may be to
humans, but they also offer many potential contributions to society. Many modern medicines are based directly or indirectly on
individual species of plants, insects, or microbes. Many species
(including trees) provide direct economic benefits. Natural species
provide the raw materials for biotechnology, a science we count on
for the future of agricultural productivity. All species are part of the
functioning global ecosystem.

Conserving biodiversity means using our landscape in such a
way as to maximize the chances for all species to survive on our
crowded planet.

In this chapter, we will first review some of the conventional
techniques for protecting and enhancing your woodland and its
wildlife habitats. These include the conservation of old-growth forest characteristics. Secondly, we will take a close look at the concept
of biodiversity, especially as it applies to North America, and discuss some of the practical things landowners can do for its conservation. Finally, we will look at opportunities for recreation and
education based on interest in natural woodland habitats.

Protecting Existing Wildlife Habitats

The science of wildlife management has determined that wildlife
has four needs in its environment: food, water, shelter, and space.
Many of the specific techniques listed here focus on providing one
of these requirements.

The most important thing you can do is to protect the variety
of natural habitats on your land. Different ecological communities
in your woodland, as well as other habitats around it, all contribute

to the variety of wildlife present. These habitats might include meadows and old fields, wetlands and ponds, and any type of watercourse. As well, a variety of stands of trees are desirable; conifers provide winter shelter for deer and rabbits, while trembling aspen may attract ruffed grouse.

You should pay particular attention to unique habitats such as bedrock outcrops, springs, meadows, or prairies. Such unusual habitats often support a significant array of otherwise rare species.

Within these habitats, protecting those trees and shrubs that provide berries and nuts for wildlife is important. Oak, beech, and walnut provide edible nuts; cherries and mountain ash provide fruit. Many types of shrubs also provide food for wildlife. Collectively, these trees and shrubs are known as "mast" species of plants. It is particularly important that you protect some mast species of trees during harvesting operations. Trees that have catkins, such as birch, willow, and poplar, are also an important food source.

Many species of wildlife depend on holes in trees, dens in the ground, or on other shelter for nesting. Such species usually depend on other species to create the shelter in the first place. Thus groundhogs excavate burrows used later by rabbits, and woodpeckers excavate cavities in trees used later by many other species of birds.

A key aspect of protecting wildlife habitats is to maintain the following nesting shelters on your land:

- any living trees with cavities or dens for wildlife
- dead snags, important for perching, food, and cavities
- fallen logs and branches, important habitats for salamanders and other species
- dens dug in the ground, with a buffer of natural vegetation around them

Pay particular attention to any habitat features that provide a variety of opportunities for wildlife to find the basic necessities of food, water, shelter, and space. These could include:

- any surface-water features, such as streams, ponds, and wetlands, including the ephemeral wetlands present in the spring but dried up in the fall; they are a critical frog habitat
- any natural corridors of vegetation, such as fence-rows or buffers along streams – the wider the better
- stands of coniferous trees that provide shelter for mammals, such as deer, in severe weather or in the winter months

Refer back to the list of special biological features on page 62 in Chapter 3 for more ideas.

Protecting Old-Growth Forest Characteristics

Just as we have created a fragmented landscape of small woodlands in much of eastern North America, we have also disturbed most of the natural habitats remaining, if only through timber harvesting and hunting. Undisturbed woodland habitats tend to have a wider range of species present and greater genetic diversity within those species, thus making another important contribution to conserving biodiversity.

As discussed in Chapter 2, recent research has summarized a number of key features that characterize old-growth, or undisturbed, forests in eastern North America. There is considerable debate over the term "old growth," particularly in the context of West Coast rainforest logging. But here in the eastern deciduous forest, the issue is simple. If we are to continue expanding our knowledge of woodland ecology in order to manage woodland ecosystems with greater care in the future, one critical need is to have examples of older, undisturbed woodlands for comparison.

Older woodlands that have not been disturbed are our best examples of healthy, ecologically diverse woodland ecosystems. They tend to be more biologically diverse than woodlands that have been disturbed or intensively managed. As we try to develop management systems that respect natural ecosystems more directly, such undisturbed benchmarks are of critical importance.

In Chapter Three (page 66), we have already listed some of the old-growth characteristics to look for. To protect or restore old-growth features representing undisturbed forests, you need only leave your woodland alone for a century or two. Given that we are not around for that length of time, some specific management options you should consider include the following:

- allowing dead trees to fall naturally and to remain as decaying logs, forming pits and mounds where they fall
- leaving branches and fallen debris to decay on the ground
- protecting cavity trees and snags for nesting wildlife
- protecting mast trees such as oak, beech, and cherry as food for wildlife
- leaving very large trees and supercanopy trees in your woodland; supercanopy trees provide roosting sites for hawks and other birds of prey

Supercanopy trees provide roosting perches and nest sites for hawks and eagles.

Protecting Waterways and Wetlands

Among the most important habitats to protect on your land are waterways and wetlands. Buffers of undisturbed natural vegetation along and around all waterways or wetlands are of high ecological value. Experts argue over the best width of such buffers, but a minimum of 50 feet (15 m) should be considered. On steep or unstable slopes, a buffer to the top of the slope is required.

In these buffers there should be little or no harvesting of trees, no skidding trails, or road construction. No heavy equipment should be used that would lead to erosion of soil, resulting in sediments entering a stream. Some states have now legislated requirements for such buffers in all forest operations, and buffers are widely used in agriculture.

Some drainage features, such as springs, seepage zones, and some wetlands, dry out later in the summer season. This does not mean they should not be protected. Fluctuations in water regimes are a normal feature of woodland ecosystems; in determining areas to protect, you should walk your woodland in early spring to make note of all wet areas. These are the areas you should protect from disturbance.

Protecting Soils and Slopes

The same comments can be made about erodable soils and steep slopes. Soils vary considerably in their tendency to erode, but any site with bare, eroding soil should be protected from further disturbance, and if necessary, rehabilitated with natural vegetation. This is particularly true for areas along watercourses, where eroding sediment can enter the stream, destroying aquatic life, including fish habitats.

Steep slopes tend to erode most easily, especially if they are composed of erodable soil types. Such slopes should be left in forest cover, with no access by heavy equipment and no attempts to build roads across them.

Fencing out cattle is one of the first steps to take in protecting your woodland.

Protecting Your Woodland from Cats, Dogs, and Cows

Wandering domestic and feral cats are the number-one predator of small birds and mammals in eastern North America. Keeping your cat near home (or inside) during the spring nesting season is a direct way to protect wildlife (except for mice and rats in the barn!). Dogs often chase and kill wildlife as well, especially if they are allowed to wander in packs.

If necessary, you should fence off your woodland to ensure that cattle do not use it for grazing. Cattle appreciate the shade and graze on young plants and saplings, destroying undergrowth and the regeneration that keeps your woodland healthy.

In severe cases of grazing, the earth inside the woodland becomes heavily compacted and all younger trees and smaller plants die, leaving only sparse taller trees. Grazed woodlots are easily recognized because you can see right through them at ground level due to the absence of lower vegetation. Eventually, cattle can kill mature trees and destroy the woodland by trampling through it.

You can create shade for cattle by the judicious planting of trees, with wide surrounding fences to protect those trees from being trampled.

Restoring Degraded Woodland

It is not always easy for the individual landowner to restore woodland habitats. Restoration of a degraded woodland may require the assessment and advice of a professional. Woodlands can be degraded through too-intensive harvesting or inappropriate selection during harvesting, through cattle grazing, through inappropriate recreational use, or through the spread of invasive non-native species.

The simple key to restoration is to protect the woodland from further disturbance and leave it alone. However, if the composition of species in your woodland has been severely altered, it may be more effective to undertake some selective management in order to encourage the regrowth of a more natural ecological community. You should consult a professional biologist or forester if you suspect that this is the case with your woodland. Harvesting can be planned to influence the species composition of the young trees that will regrow after cutting.

Enhancing Wildlife Habitats

If wildlife habitats are lacking on your land, there are many things you can do to enhance them. These range from planting cover and food plants to building brush piles and putting up birdhouses.

You can start by planting trees or shrubs that will provide food sources or cover for wildlife. Trees and shrubs that are important food sources for wildlife are shown in the following table. It is now widely recommended that when doing this you always use species native to the region where you live; some non-native species, such as Norway maple or purple loosestrife, become invasive and destroy wildlife habitats.

Table 12   Trees and shrubs for wildlife

SHRUB SPECIES	TREE SPECIES
Common elder (Sambucus canadensis)	*American beech* (Fagus grandifolia)
Dogwood species (Cornus spp.)	*Cherry species* (Prunus spp.)
Highbush cranberry (Viburnum trilobum)	*Hop hornbeam* (Ostrya virginiana)
Nannyberry (Virburnum lentago)	*Maple species* (Acer spp.)
Serviceberry (Amelanchier spp.)	*Mountain ash* (Sorbus americana)
Staghorn sumac (Rhus typhina)	*Oak species* (Quercus spp.)
Raspberry (Rubus spp.)	*White birch* (Betula papyrifera)
Riverbank grape (Vitis spp.)	*White cedar* (Thuja occidentalis)
Trumpet honeysuckle (Lonicera sempervirens)	*White pine* (Pinus strobus)
Virginia creeper (Parthenocissus)	*White spruce* (Picea glauca)

Source: J.M. Daigle, and D. Havinga. *Restoring Nature's Place.* Schomberg, ON: Ecological Outlook Consulting, 1996.
C.L. Henderson, *Landscaping for Wildlife.* Minneapolis: Minnesota Dept. of Natural Resources, 1987.

Nesting cover and shelter can also be improved on your land. Many small mammals, reptiles, and amphibians will benefit from brush or rock piles and from fallen logs left to rot on the forest floor. If these are not naturally present in your woods, you can create them. When thinning your woodlot, leave some logs to rot or leave some larger logs that are unacceptable as sawlogs on the forest floor.

To build a brush pile or stone piles, place larger logs or rocks on the bottom before covering them with smaller branches or stones. This will create permanent cavities underneath that wildlife need for protection.

If you have few trees with cavities for birds to nest in, you can, of course, make and erect birdhouses of all kinds, for species ranging from bluebirds to wood ducks.

You can even create dead snags that will provide food sources and encourage cavity nesters, such as woodpeckers, by girdling a few trees in your woodlot. One easy way is to girdle some trees when thinning your woodlot; this is a lot less work than cutting them for firewood.

Step 1 – The foundation

Step 2 – Build piles

Step 3 – Finished brushpile

Building brush piles provides habitat for many mammal species. Be sure to place large logs or rocks on the bottom to create permanent hollow spaces.

Table 13   Sample birdhouse sizes

BIRD	ENTRANCE DIAMETER*	ENTRANCE HEIGHT	FLOOR DIMENSIONS
American kestrel	3	10–12	8 x 8
Barred owl	6–8	14–18	13 x 13
Black-capped chickadee	1.1–1.5	6–7	4 x 4 to 5 x 5
Eastern bluebird	1.5	6–7	4 x 4
Northern flicker	2–3	10–20	6 x 6 to 8 x 8
Nutchatches	1.1–1.5	6–7	4 x 4 to 5 x 5
Tree swallow	1.25–1.5	6–7	4 x 4 to 5 x 5
Wood duck	3 x 4	16–18	10–10 to 12 x 12

* All dimensions given in inches.

Source: D. and L. Stokes. *The Complete Birdhouse Book.* Boston: Little, Brown, 1990.

Conserving Biodiversity

The forests that developed in the several thousand years since the last glaciation in eastern North America tended to contain large,

Erecting birdhouses can be a family activity which provides improved habitat for selected species.

unbroken canopies of trees. The settlement clearings carved by native peoples, the long tornado tracks of downed timber, and the occasional patches of prairie pushing eastward were still relatively small openings in an unbroken expanse of forest.

It is not surprising that the wildlife species found in this region included many that are adapted to living in the forest interior. Of these species, birds have been studied most intensively. Known by biologists as forest-interior species, such birds now find their habitat in short supply.

On the other hand, forest-edge habitats have multiplied dramatically. As we clear the landscape for agriculture and increasingly for urban development, we have left behind a very fragmented forest, consisting mostly of small patches of woodlands with some fencerows and wetlands. These small areas of natural vegetation provide an excess of edges, known to be good habitat for other, generally more common, species of birds and mammals.

Not surprisingly, the populations of these forest-edge species have been expanding for many years and find no shortage of appropriate habitats. (It should be noted that some specialized habitats, such as open undisturbed meadows, prairies, and oak savannahs, are now extremely rare, having been almost completely cleared for agriculture.)

To keep the explanation simple, the central message of conservation biology is that we need to restore this balance somewhat, particularly by maintaining and expanding the number of large woodlands and by maintaining or creating corridors of natural vegetation between them. Current scientific research suggests that we will conserve biodiversity most effectively if natural habitats are large and are not isolated from other natural habitats.

Science suggests that two of the most important practical strategies for a landowner interested in conservation are the following:

1. Expand your woodland through reforestation or by allowing natural succession to create as large a patch of woodland as possible; solid blocks of woodland provide more forest-interior habitat than elongated narrow woodlands.

2. Connect smaller patches of woodland or connect your own woodland to other woodlands with corridors of

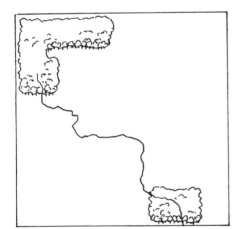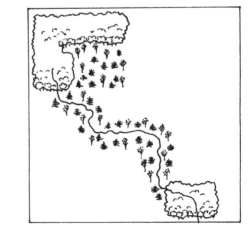

Solid blocks of forest and connected woodland patches provide for improved conservation of genetic diversity of both plants and wildlife.

natural vegetation; such corridors provide double benefits if they also serve as natural buffers along waterways.

Implementing such a strategy beyond your own property might even lead to cooperation among neighbors in maintaining a particularly large natural woodland. Some neighbors go beyond this to share their woodlands for recreation with a common walking trail. As long as you leave the woodland alone during the spring nesting season, such a trail will not likely disturb the wildlife.

In addition to keeping your woodland as large as possible and connecting it to other woodland patches, conserving biodiversity may require giving special attention to species that are vulnerable in a variety of ways. This takes us back to the question of rare species and their management.

It is usually beyond the ability of landowners to identify rare species on their own land, especially rare species of plants. However, state or provincial agencies, as well as local biologists, should be able to help you evaluate your land from this perspective. If you find significant rare species, you should adapt your management plans to protect them.

Other practical management options that help to conserve biodiversity are discussed elsewhere in this book; some of the most important techniques are

- protecting or restoring buffers along watercourses,
- protecting all unique habitats, including wetlands, and
- using native plants for reforestation.

Education and Recreation

Landowners who are interested in wildlife, bird-watching, or in the identification of native plants have enormous opportunities to gain a deeper appreciation of their woodland ecosystems. You can go far beyond the basic inventory of your woodland to keep notes and records of species seen and of changes over the seasons. Going for a regular walk through your woods opens your eyes to the living natural history being played out in front of you.

Keeping a journal is one way of deepening your appreciation of nature in your woodland. If you keep recording your observations of woodland visits over the years, you can accumulate a fascinating record reflecting the seasonal rhythms of nature. You may find regular events – such as deer tracks in December, spring wildflowers, foraging porcupines in May, and mushrooms in September – that help mark the year. Such accumulated observations provide a rich store of knowledge that will, in the long run, add to your enjoyment of your woodlot and probably to your commitment to conserve it.

Sharing nature with children is a gift you can give to the next generation. Time your woodland walks to avoid the late spring and summer bugs, but take your children into the woods in the early spring or in the fall and help them learn to appreciate the beauty of nature. The fact my mother loved to go for a walk in the woods is one of the main reasons that I have ended up with a similar love.

This basic appreciation of nature is reflected in the final chapter of this book, a summary of our observations over the seasons. Read it as a guide to what you might see in your own woodland.

Volunteer-Monitoring Programs

You can go beyond a personal love of nature by participating in regular volunteer-monitoring programs, counting birds, frogs, or other species. This may take you beyond your own property to visit other natural areas, or it may simply mean keeping a record of birds at your own feeder. State and provincial wildlife offices can tell you about opportunities in your own region. Participating in such volunteer programs is a useful contribution to expanding our ecological understanding, but it also benefits you personally as you learn more natural history.

You may wish to go beyond appreciating nature yourself and join a local natural-history society. Such groups of amateur naturalists meet to share their observations, to hear speakers on numerous topics of interest, and to participate in nature walks or outings. Some get actively involved in ecological-restoration projects. Some starting points for finding information on such opportunities are provided in the appendix.

Beyond studying nature in your woodland, you may wish to invite a professional biologist or a university class to conduct a more detailed inventory, identifying the plants and wildlife that you do not know or simply walking through your woodland with you to help expand your knowledge. This may be particularly important if you feel that rare species may be present in your woodland.

Nature Trails

The simplest form of recreation is going for a walk in your woods, but many landowners may be interested in more active recreational pursuits, such as cross-country skiing or hunting. Clearing a trail through your woods can open the door to these activities, as well as to regular nature walks. The design and building of trails is discussed in Chapter 9.

Leaving woodlands alone is almost always a viable management option, especially for landowners who want to emphasize biodiversity and the conservation of wildlife habitats. But there are many practical things landowners can do to improve environmental sustainability and wildlife habitats while at the same time deepening their own understanding of and relationship with their woodlands.

These improvements can be done alone or in combination with a plan to harvest timber and firewood, as described in the next chapter.

Further Reading

Agriculture Canada. *Best Management Practices: Fish and Wildlife Habitat Management.* Guelph, ON: Agriculture Canada, 1996.

Cox, Jeff. *Landscaping with Nature.* Emmaus, PA: Rodale Press, 1991.

Decker, Daniel J., and John W. Kelley. *Enhancement of Wildlife Habitat on Private Lands.* Ithaca, NY: Cornell University Extension Service, 1986.

Decker, Daniel J., et al. *Wildlife and Timber from Private Lands: A Landowner's Guide to Planning.* Ithaca, NY: Cornell University Extension Service, 1983.

Hassinger, Jerry, et al. *Woodlands and Wildlife.* University Park, PA: Pennsylvania State University, 1979.

Henderson, C.L. *Landscaping for Wildlife.* St. Paul, MN: Minnesota Department of Natural Resources, 1986.

Hunter, Malcolm L. *Wildlife, Forests and Forestry: Principles of Managing Forests for Biological Diversity.* New York: Prentice Hall, 1990.

Landowner Resource Centre. "The Old-growth Forests of Southern Ontario" (fact sheet). Manotick, ON: Landowner Resource Centre, 1996.

———. "Restoring Old-Growth Features to Managed Forests in Southern Ontario" (fact sheet). Manotick, ON: Landowner Resource Centre, 1996.

Yahner, Richard H. *Eastern Deciduous Forest: Ecology and Wildlife Conservation.* Minneapolis: University of Minnesota Press, 1995.

CHAPTER 6

Timber and Firewood Harvest — Principles and Practices

Whether you intend to manage your woodland primarily for timber and firewood production or not, you can benefit by understanding the techniques of proper management for this purpose. You may simply want to harvest a few sawlogs on occasion or cut a little firewood for your own use. Or you may plan an ongoing regular timber harvest. In either case, you can maximize both your economic return and the sustainability of the environment through better understanding.

Even if you never intend to harvest any timber, it is worthwhile to be aware of the economic benefits. Growing trees is like having money in the bank. You might think of your woods as a retirement savings plan, in case you ever need it. With changing tax laws, it may be that the only way your children can afford to inherit your woodland will be through a one-time timber harvest, unless you have other wealth with which to pay capital gains or estate taxes. You can think of continuing good management as insurance against this possibility.

If your woodland has been degraded in the past through cattle grazing or too-intensive logging, you may also be able to improve biodiversity and wildlife habitats through some cutting or thinning.

Forest Types and Forest Products

As noted in Chapter 4, there are several different forest types you may find on your land, as well as different associations of trees Most timber management is undertaken in upland deciduous or mixed woodlots consisting of hardwood trees. This is where the financial return is greatest. The products are sawlogs and veneer logs, as well as firewood.

Other products might include

- utility poles, barn poles, or landscape squares from coniferous plantations,
- fence posts from cedar swamps,
- pulpwood or chips from coniferous forests, and in some areas, from poplar stands,
- firewood and pallets from soft maple swamps, and
- specialty products, such as burls for carving, ash or cedar for canes, and hickory for tool handles.

The case of managing coniferous plantations will be considered in Chapter 7, but there is little or no literature on woodlot management for these other products (except for pulpwood in northern forests) and little in the way of specialized management to be undertaken.

Specialty products depend heavily on local markets; your provincial or state woodlot-owners' association can lead you to information on possible alternate markets. It is up to you as a landowner to be aware of the possibilities and pursue these locally if you are interested.

This chapter focuses on the management required to harvest timber and firewood in a hardwood forest; this is the economic mainstay of the forest industry in eastern North America.

Historical Background

Timber-harvesting practices in North America have gone through several revolutions. Early practices were totally exploitive. With thousands of miles of forest stretching west, early white settlers worried little about running out of trees. But even within a generation or two, the landscape had been almost totally cleared in certain regions, and firewood was in short supply. Illustrations of the settled landscape in eastern North America in the mid-1800s show a countryside largely devoid of trees, except in the remnant woodlots behind the farms. Certainly the landscape had far fewer mature trees in fencerows than it has today.

By the late nineteenth century, certain individuals were becoming concerned about the lack of woodlands. Disappearing woodlands were related to increases in spring flooding, and knowledgeable citizens began to argue for conserving forests and replanting trees. Arbor Day started in the late 1800s as a way to encourage citizens to plant trees in urban areas. One of the early arguments supporting this was that trees were the "lungs of the city," based on an understanding of photosynthesis. Trees use carbon dioxide and release oxygen, thus contributing to improved air quality.

During the first half of the twentieth century, we gradually learned to use more sustainable forestry practices, at least in terms of reforesting areas that had been harvested. Government agencies started tree nurseries and encouraged reforestation of eroded soils as well. But such forestry has not been sustainable in the modern sense of the term, since reforestation was often unsuccessful and frequently replaced a natural ecosystem with a single-species plantation.

Today we are learning the benefits of other forest values such as water conservation, wildlife habitat, and biodiversity. Forest management takes the ecosystem approach we have discussed in this book, taking all values into account. Modern ecologically based timber or firewood harvests should mimic the natural ecosystem, while protecting the other environmental values of the woodlands.

The Theory of Management for Timber and Firewood

One simple concept underlies much of the management undertaken for timber in the hardwood forests of eastern North America. This is the concept of stocking.

"Stocking" refers to how crowded or uncrowded the trees in your woodland are. If trees are too crowded, they will grow more slowly, competing with each other for available sunlight, nutrients, and moisture. If trees are not crowded enough, they will grow rapidly, but they will develop extensive lower branches reaching sideways, leaving them with a short main trunk and thus unuseable for timber.

In the most general sense, management involves thinning overstocked (overcrowded) stands to achieve more growth or leaving understocked (undercrowded) stands to grow until they achieve adequate density. In fact, such management can be used to grow larger trees, whether or not you intend to harvest them for timber. The trick is in knowing exactly how much to thin your woodland and how to pick which trees to cut.

Sustainable timber or firewood harvesting in a woodland should only remove enough trees to maintain appropriate stocking levels, while respecting all other woodland values and taking into account the higher economic return from older, larger sawlogs. This will maximize long-term economic return, as well as environmental sustainability.

The most common mistake in harvesting timber is to cut too many trees and to cut trees that are still small sawlogs, having just made the minimum size needed by sawmills. You are far better off both ecologically and economically to cut fewer trees, allowing them to grow to a larger size before harvest. Trees gain the fastest increase in volume after reaching the minimum sawlog size, providing their greatest economic return as they grow to larger size classes.

The most common mistake in cutting firewood is to cut larger or dead trees for firewood. Larger trees bring a much higher economic return if sold as sawlogs. Dead trees no longer compete with others for nutrients or moisture and are more valuable as wildlife habitat. It is usually better to cut medium-size trees for firewood, as part of a planned woodland thinning, than to cut larger trees, unless the large trees are of low quality, diseased, or of no value as den trees for wildlife. You can also cut firewood from the tops of trees harvested for timber.

It is not possible in this book to provide a detailed technical explanation of all appropriate timber-harvesting options for all the woodland types in eastern North America. This chapter will help you gain an understanding of the concepts involved; you should consult a professional forester before planning any active management or harvesting in your woodland, particularly to ensure that you are using the correct local guidelines for optimum stocking.

Thinning in a coniferous plantation is discussed in Chapter 9. The concepts presented in this chapter apply most directly to hardwood forests, which are usually dominated by sugar maple, beech, hemlock, white ash, black cherry, oak, and hickory.

Steps in Timber and Firewood Management

1. Preparing an Inventory

Preparation of woodland inventories has been discussed in Chapters 3 and 4. The first step in managing woodland for timber harvesting is to prepare a detailed inventory of it. The inventory should identify the "ecological communities" or "forest compartments," which are stands of similar trees, that you will manage in a standard fashion. For each of these forest communities or compartments, you need to prepare a timber inventory, including the size and species of trees present. You may wish to hire a professional forester to do this, or at least to check your results.

Your inventory should include all other factors such as soil conditions, drainage patterns, rare species, and wildlife habitats. These are important considerations to take into account first, to ensure that your timber-management plans meet the basic minimum level of environmental sustainability.

Timber-management decisions are based on the information you gather and analyze in your timber inventory.

2. Evaluating Environmental Sensitivity and Selecting Crop Trees

The first step after preparing your inventory is to establish areas of environmental sensitivity that you will leave alone and not manage for timber or firewood production. As discussed in Chapter 5, you should first protect biodiversity, water resources, and wildlife habitat to sustain the woodland ecosystem.

The second step is to select desirable crop trees for possible harvest. Crop trees should be those of desirable species, with the following characteristics:

- healthy, tall, and straight
- few or no branches below the crown
- vigorous crown one-third the height of the tree
- dominant position in the canopy

These are the trees which you will protect during any thinning, firewood harvest, or during other operations, until they grow to the size when you will harvest them. Even if you are not planning to harvest timber, but are treating your trees as money in the bank, these are the trees you should protect and conserve. If they do not make eventual sawlogs, they will make beautiful old-growth trees. The highest-quality trees among your crop trees – those with few if any knots and very straight trunks – may be graded as veneer logs, which bring a substantially higher price. All trees will be graded during a timber harvest, and prices will vary accordingly.

The tallest and straightest trees should be selected and protected as potential crop trees.

Although you may know the trees in your woodlot quite well by now, it may help you when choosing crop trees to consider the information in Tables 2 and 11 (pages 75 and 92), which list the values of trees for timber and firewood.

Trees that are crooked, diseased, or otherwise damaged will never make useful sawlogs. Depending on their size, they could be harvested for firewood, left as den trees for wildlife, or left to provide enough stocking to encourage the upward growth of other trees. You should always be careful not to cut too many trees, in order to preserve the stocking in your woodland at a desirable level.

3. Planning a Thinning or Improvement Cut

Thinning and improvement cuts, as well as timber harvests of mature trees, will influence the future species composition of your woodlot, depending on how they are undertaken. The planning is the same; whether you gain a harvest of timber in the end depends on the size-class distribution of the trees. This is an important

Thinning means to remove a number of poorer quality and crowded trees to allow improved growth of the best trees.

point: you do not plan a timber harvest in advance; you evaluate the stocking of your woodlot to determine whether or not a harvest is appropriate at this time. You will often discover a need to thin some trees that are below sawlog size, and therefore destined for firewood, as well as possible sawlogs.

- A thinning cut is done in stands greater than 4 inches (10 cm) in diameter, to give crop trees room to grow.
- A cleaning cut is done in younger stands to influence species composition, if a woodlot has previously been degraded.
- An improvement cut is done in older stands that have been poorly managed or grazed in the past.

In order to understand the planning necessary to thin your woodland, we return to the concept of stocking. With a detailed timber inventory, you can develop a thinning plan based on calculations of tree density and size. It is also important to know whether your woodland is made up of an "even-aged" or "uneven-aged" stand of trees.

Even-Aged or Uneven-Aged? In the hardwood forest of eastern North America, the most appropriate management is usually one that aims to create or maintain an uneven-aged stand of trees; that is, there are trees of all ages (and sizes) in the woodland. In this case, there will be many smaller trees and smaller numbers of larger trees. But some woodlands are composed primarily of trees approximately the same age. They may have grown up after old fields were abandoned or after intensive logging or clear-cutting.

You can tell whether your woodland tends to be even-aged or uneven-aged by comparing the following two graphs. In the first case, most trees are clustered around an even age of forty to sixty years, with a few trees older and younger than this. This is referred to as even-aged (it does not mean that all trees are exactly the same age). In the second case, there are many young trees and less of each size class until there are only a few of the largest trees. This is referred to as uneven-aged. In this case, there are plentiful young trees to replace those being harvested.

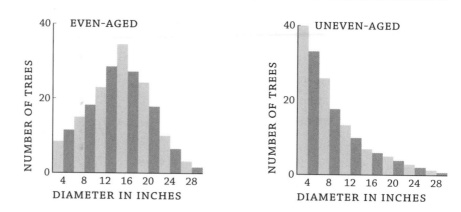

Even-aged and uneven-aged forest stands: Uneven-aged stands tend to have a large number of smaller trees, and fewer numbers of larger trees, as shown in the right graph; even-aged stands tend to have a predominance of trees clustered around a few middle-aged size classes, as shown on the left. (Numbers conceptual only.)

Source: adapted from P. Williams and S. Pease. *An introduction to Farm Forestry.* Guelph, ON: Dept. of Environmental Biology, University of Guelph, 1993.

In sustainable timber harvesting in the hardwood forest, you aim to have an uneven-aged stand and to harvest the maturing timber as it reaches the largest economic size, while leaving plentiful younger trees to continue growing. If your stand is even-aged, you should consult a professional forester to guide you toward management options that will gradually transform your woodland into a more uneven-aged condition. Recommendations for even-aged stands and shade-intolerant species are difficult to generalize; they depend on site-specific conditions.

CALCULATING STOCKING To calculate stocking, a forester counts the number of trees per acre and measures their diameter (see Chapter 5). Using the diameter measurements, the basal area of each tree is calculated, and a total basal area per acre is established. You usually count only trees greater than 4 inches (10 cm) in diameter in this calculation.

Using the diameter measurements from your sample, you can first establish whether your woodland tends to be even-aged or uneven-aged. Count the number of trees you have found in each of the diameter classes, using the above diagrams as a guide; you can portray this visually by constructing a graph for your own woods.

Woodland walks in all seasons provide the best way to start learning about and understanding forest ecosystems. The magic of the woods is revealed in its cycles – cycles of life and death, of growth and decay, and of changes through the seasons. The changing colours of the woods, from the bright green of spring through the yellows and orange of fall to the stark white of winter are merely the most obvious pattern of change. Blooming wildflowers and migrating birds make spring unique. By early summer the woods is quiet while birds raise, feed and protect their young. Fall colours make another magic transformation in October, caused by a complex chemical reaction to changing temperatures. Winter snow suddenly reveals the tracks of animal residents that are actually there all year round.

Soil, water and the history of land use are important influences on your woodland. Low wet areas or forested swamps (top right) need to be protected from disturbance. If your woodland is growing on shallow rocky soil (center right), its growth will always be somewhat limited. An old cedar fence line found in the middle of a woods may tell the story of a former pasture (bottom).

The line of old maples (background) with their branches reaching sideways for the sun represent an old woodlot edge; the young trees on the right are growing on former pasture.

Learning to identify trees by their leaves is the first step in describing your woodland. Golden beech leaves (top left) appear alternately on twigs and have distinct patterns of veins; teeth along the edge of the bright green elm leaves (middle left) have a distinctive 'double-toothed' pattern, with some teeth consistently larger than others. Bright red maple leaves are easily recognizable; they always appear exactly opposite one another on the branch. Other characteristics like the heavily ridged bark of this white ash (background) can also help identify individual tree species.

A healthy woodland ecosystem contains hundreds of other species besides trees. Like many spring wildflowers, the spring beauty (lower right) is adapted to take advantage of early spring sunshine, blooming before the canopy of tree leaves shades out the sun's energy. The graceful fronds of maidenhair fern (center) are adapted to grow in the shade of the canopy, their flat green leaves capturing as much light as possible. The scarlet tanager (top) is but one of many woodland bird species. Uncommon species such as this towering tulip tree (background) also add to biodiversity.

Woodland ponds and wetlands, as well as woodland streams, all help sustain water quality and add to biodiversity. Timber harvests and trails should avoid these areas as much as possible.

Properly constructed woodland roads (right) can make walking a pleasure, and will help minimize environmental damage if timber is harvested.

Dead trees or 'snags', rotting stumps, brush piles and logs lying on the forest floor all provide important habitat. These are the home for thousands of insects and small amphibians, like salamanders, a largely invisible, but critical link in forest food chains. Resist the urge to 'tidy up' the fallen branches or dead trees; they are a natural part of the forest ecosystem.

Woodpeckers are forest engineers, excavating feeding cavities like these (upper right), or nesting cavities later used by numerous other bird species. Brushpiles (background, right) can provide extra habitat for small mammals like rabbits.

The aim of management in hardwood forests is to create a forest that has trees of all ages and sizes (inset right). To manage your woods for timber or firewood production, the tallest, straightest trees should always be saved as possible 'crop' trees (background, this page). This can begin when younger stands of trees reach an average 10 cm. (4 inches) diameter. Protect the straightest, and thin some of the poorer trees (upper left); those remaining will grow more quickly. Thin closely grouped trees by cutting the mishapen ones, always leaving the straighter, larger trees. In photo (lower left) the center tree would be kept for timber and the ones on the left and right cut for firewood. The woodlot on the right (background) was significantly thinned two years before this picture was taken.

Firewood can be obtained by cutting larger trees that are not straight enough to ever be harvested for timber.

Protect diversity in your woodland by keeping some evergreen trees such as the dark green hemlock partially obscured at the left of this photo. Trees that are especially valuable, such as the black cherry (inset photo, this page, recognizable by its bark – said to resemble large burnt corn flakes), also deserve special protection for future harvest.

Enlarging your woodlot, and planting corridors that connect your woodlot to other nearby woodlands can be a major contribution to protecting biodiversity. Woodlots can expand through the 'natural succession' of plants that will grow when a field is left alone, through reforestation, or through a combination of these. Coniferous trees such as white pine (inset, above) compete well against weeds and grasses, and are therefore widely used for reforestation. Deciduous trees (left) require tender loving care – some mulch and a tree guard will be necessary to get them started.

Mowing a walking trail through old fields before they grow up in trees, or before reforestation, can provide easy recreation. You can do this with any lawnmower, and then the trail is great for walking or cross-country skiing.

Getting your children involved in making maple syrup is just one of many ways to introduce them to the beauty of the woods.

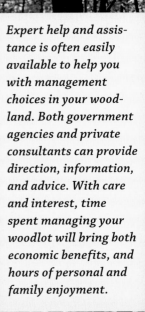

Expert help and assistance is often easily available to help you with management choices in your woodland. Both government agencies and private consultants can provide direction, information, and advice. With care and interest, time spent managing your woodlot will bring both economic benefits, and hours of personal and family enjoyment.

If your sample tends to have many younger trees and fewer older ones, it will look somewhat similar to the uneven-aged stand illustrated above. If the trees in your sample cluster around a middle-size or age range, it will tend to look similar to the even-aged stand also illustrated above.

To calculate the level of stocking in order to establish whether some timber harvest is appropriate, add up the total basal area per acre. Here it pays to consult a forester, because the total basal area could be composed of many small trees or a few large ones; some judgment on management options is inevitably involved. However, optimum stocking after harvest in a northern hardwood forest stand should consist of 60 to 70 square feet per acre (14–16 square meters per hectare) in trees greater than 9.5 inches (24 cm) in diameter, depending on your exact location.

If your stocking level is above the maximum recommended, your woodlot is overcrowded and can be thinned or harvested down to the optimum by removing selected trees. If your stocking level is between maximum and minimum, the trees should be left to grow. If it is below minimum, your woods should be left to grow as well, but might also benefit from special treatment to reach the desired stocking level.

You can also get a visual impression of whether or not your woodlot should be thinned by looking at the canopy. A completely closed canopy with no gaps of light showing between trees probably reflects a need for thinning. A correctly thinned woodlot will have gaps of at least a few feet around at least two sides of the crowns of crop trees. An under-stocked woodlot or a woodlot that has been cut too intensively, will have major canopy gaps between trees. This may still be useful forest management, depending on the species you wish to regenerate.

If your woods needs thinning based on these calculations, the final challenge is to choose which trees to thin. For this go back to your original selection of crop trees, and keep the best, tallest, and straightest trees, while taking out enough of the slightly smaller, crowded, poorly formed trees to achieve the desired stocking level.

MANAGEMENT OF UNEVEN-AGED WOODLOTS If your woodland tends to be uneven-aged, there are two rules of thumb that you can use to judge which trees and how many to cut. If the stand averages over 6 inches (15 cm) in diameter, crop trees should be spaced "D" plus 6 feet (1.8 m) from the next crop tree, where "D" is the tree's diameter in inches with each inch representing one foot. Therefore, a 10-inch (25 cm) crop tree should be spaced (10 + 6 =) 16 feet (5 m) from the nearest large tree. If the stand averages less than 6 inches (15 cm) in diameter, use "D" plus 4 for your calculations. However, tree quality is more important than spacing; keep high-quality trees even if they are close together.

The second rule of thumb is to allow the crown of the tree enough space to equal twenty times the diameter of the tree. Thus, a 10-inch (25 cm) tree would require crown spacing of (10 inches x 20 = 200 inches =) 16 feet (5 m). A thinning cut to release crop trees for faster growth should open up the crowns of crop trees on at least two sides.

When picking trees for harvest, you should follow a plan to leave a balanced range of trees of all sizes after cutting (what

Table 14   Recommended residual stand stocking

DIAMETER CLASS	NUMBER OF TREES	BASAL AREA IN SQUARE FEET	NUMBER OF TREES	BASAL AREA IN SQUARE METERS
	per acre	per acre	per hectare	per hectare
5–9 inches (12–24 cm)	65	16	160	3.7
10–14 inches (26–36 cm)	28	22	69	5.1
15–19 inches (38–48 cm)	17	26	42	6.0
20 inches or greater (50 cm or greater)	8	20	20	4.6

Source: C. Arbogast. *Basic Principles of Forest Management in* Northern Hardwoods. Land O'Lakes, WI: Northern Hemlock and Hardwood Manufacturer's Assoc. 1956.

J. Irwin. *So You Own A Woodlot: Now What?* Cambridge, ON: Ministry of Natural Resources. 1996.

foresters call "residual stocking" or "stand structure"). The following table is one guide for the basal area of trees that should remain in four different size classes after harvesting.

To apply these guidelines to your own woodland, consult a local professional forester to find the locally used standards. In the eastern deciduous forest region these vary with forest type and location.

The same concept can be expressed another way, by comparing the distribution of tree sizes in your woodland to an ideal distribution. Consider the following figure which illustrates an ideal range of tree sizes compared to the number of trees in each size class in an actual woodlot. When you prepare the inventory of your woodlot, you may discover that you have more or fewer trees in any one size class than the ideal.

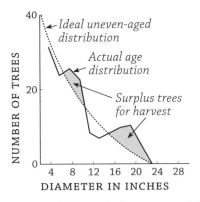

Size distribution chart: This graph shows ideal uneven-aged distribution compared to actual distribution of tree size classes – where actual exceeds the ideal, "surplus" trees are available for harvest; these could be in any size class. Consult a forester for actual guidelines. (Numbers conceptual only.)

Source: adapted from H.W. Anderson, et al. *A Silvicultural Guide for the Tolerant Hardwoods Working Group in Ontario.* Toronto: Ministry of Natural Resources, 1990

The idea of this diagram is that the trees that are outside the ideal distribution are available for harvest; cutting them maintains the uneven-aged distribution of trees in your woodland, thus optimizing overall growth. In other age classes, your woodlot may be short of trees, and there should be no harvesting. Note that if you follow this model, you may be harvesting trees of several different sizes, some for sawlogs and some for firewood. Cutting all the trees over a certain diameter would be "high-grading" your woodlot, leaving a poorer stand for future growth.

We have deliberately left this diagram very general; a forester can assist you in selecting trees in your own woodland to maintain or improve the stand structure after harvesting.

Unfortunately, this ideal way to plan a timber harvest is often contradicted by local bylaws. These may only stipulate that you cannot cut trees smaller than a certain diameter, such as 16 inches (40 cm). A "diameter- limit cut" is a timber harvest that cuts everything above this diameter limit, obeying the bylaw, but practising very poor forestry.

There are two main problems with a diameter-limit cut, both of which can cost the landowner a substantial share of their economic return. First, cutting everything above a set diameter may include many trees that are just beginning their period of optimum growth and highest economic return; this means that the landowner gets the minimum economic return.

Secondly, a diameter-limit cut with no reference to stocking rates may open up a woodlot to such an extent that regeneration does not result in a growing stock of trees for harvest for several decades. Initial regeneration may be of undesired species and may establish open-growth forms that make them unsuitable as sawlogs. Not until a canopy is reestablished will a good stocking rate be in place.

Another common problem even in carefully managed woodlots is that the tallest, straightest trees are eventually cut for their timber value. This is, after all, what timber management is all about. However, a few of these trees should be kept as seed sources for future generations. When only the lower-quality trees are left, we breed future generations of poorer-quality trees. As one forester said to me recently, this is like shooting all the winners of the horse race and putting the last-place finisher out to stud. By keeping a few of the best trees in your woodlot as seed sources, the future genetic variability within the trees of one species may make a significant difference in their growth.

The correct way to calculate and mark timber for harvest is to measure present stocking and compare it to an ideal; then cut trees out of all size classes (not just some of the larger trees) to bring the residual stocking as close to ideal as possible.

MANAGEMENT OF EVEN-AGED WOODLOTS If your woodland tends to be even-aged, it is even more important to consult a forester; management recommendations depend heavily on site-specific conditions. One useful rule of thumb is to select crop trees approximately every 22 to 23 feet (7 m) in all directions; protect these and thin some of the trees in between, cutting those that are of lower quality, smaller, or crowded. The product of a first thinning cut is often mostly firewood.

Crop trees to be protected for future growth could be from 10 to 30 feet (4–10 m) apart; the quality of the tree is more important than the spacing. In all cases, crop trees should be those that are tall and straight, with crowns that occupy a dominant position in the canopy.

Foresters use stocking diagrams to assist in deciding how much thinning or harvesting should occur in an even-aged stand. An example is provided below.

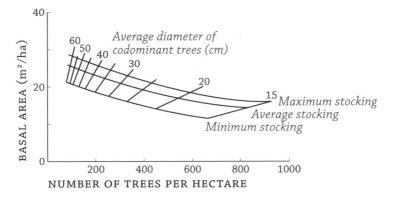

Sample stocking diagram: diagrams like this assist foresters in making specific recommendations for management of even-aged hardwood forests stands.

Source: H.W. Anderson, et al. *A Silvicultural Guide for the Tolerant Hardwoods Working Group in Ontario.* Toronto: Ministry of Natural Resources, 1990

In this sample diagram, establish the position of your own woodland using three measurements: the average diameter of codominant trees (those with crowns in the canopy), the total number of trees per hectare, and the basal area per hectare. If your woodlot ranks above the maximum stocking line, harvesting is appropriate; if it ranks much lower, it should be left to grow.

Once trees reach an average of 10 to 15 inches (25–35 cm), they are nearing sawlog size and may not benefit from thinning. Consult a professional forester for management advice if your woodlot consists of trees of this size.

Fine-branched species such as black cherry and yellow birch can be thinned when their crowns begin to touch, usually at about 6 inches (15 cm) in diameter. Coarse-branched species such as maple, oak, and ash should not be thinned until they reach an average diameter of 8 inches (20 cm), to prevent the growth of extra lower branches.

If you have found this discussion of stocking levels, even-aged stands, and uneven-aged stands confusing, consult a professional forester for advice. Professionals should be able to explain your options clearly and evaluate your woodland in more detail.

Harvesting Options

Planning for a timber harvest should follow the same rationale used in planning to thin your woodland. With the advice of a professional forester, you should look at stocking levels and establish whether there is a surplus that you can afford to harvest. It is only this surplus, or level of over-stocking, that you should ever harvest. Do not harvest trees only because they have reached the minimum merchantable sawlog size.

Comparative values of harvested sawlogs show clearly that you are better off economically to wait for harvesting until your trees grow larger – if you can resist the temptation of a small economic return in the short run in favor of a larger return in the long run.

Trees increase more quickly in value as they reach larger sizes because larger sawlogs are worth a higher price per board foot, though this is gradually offset by declining growth rates.

The following is one example cited in the booklet "Timber Management for Small Woodlands": "As a hardwood tree increases in diameter from 14 to 24 inches (36-61 cm), it may nearly double in height, increase 4.5 times in volume (135 to 630 board feet), and increase tenfold in value." (Goff, et al, 1983, 24)

This is the point at which woodlot management that makes sense ecologically also makes sense economically. Leaving your trees to grow larger takes advantage of natural growth rates and will actually bring a higher economic return in the long run. Not only the value per tree increases, but the value per board foot increases, especially if the tree is of veneer quality. This is because a larger tree puts on more new wood in a year than a smaller tree, and a larger log considerably reduces labor and handling time at the sawmill.

Tree Marking

When you consider harvesting your timber, get the trees individually marked by a professional forester (not a logging-company representative). You should be aware that there are three common silvicultural systems for harvesting trees that foresters in eastern North America usually consider.

1. Selection cutting: Used in most well-managed hardwood forests, selective cutting involves harvesting individually selected trees or small groups of trees. A light selective cut will favor shade-tolerant trees such as maples and beech in the regeneration; a heavier cut may enable more intolerant species such as white ash, black cherry, or oak to regenerate.

2. Shelter-wood cutting: May be used in special circumstances such as plantation management. It involves

cutting a significant portion of mature trees, leaving others to provide shelter for regenerating saplings, and removing the balance of the stand, often in stages, after satisfactory regeneration is achieved.

3. Clear-cutting: Is used more frequently in other ecosystems, such as jack-pine stands in the boreal forest, but is rarely appropriate in this region, except in very special circumstances, such as for the replanting of hybrid poplar stands for pulpwood or for stand conversion to other species, such as the removal of Scots pine.

In most cases, private landowners in eastern North America choose to selectively harvest individually marked trees in their woodlands. It is always advisable to have a professional forester prepare or review timber-harvesting plans and to mark individual trees for harvest. It is critical that you do not harvest too extensively, leaving your woodland under-stocked with younger trees.

There are standard methods for marking that any trained forester will follow, although these procedures differ somewhat in different locations. Some jurisdictions now train certified tree markers who can carry out this work. In any case, having your woodlot professionally marked will protect your future investment in woodlot management, while optimizing your present economic return.

Timber-Volume Calculations

When your woodlot is marked for timber harvest, data will be gathered in a similar way to your original woodlot inventory, but the information will be on all marked trees, not just a small sample of the woodlot. The data will include diameters and heights of sawlogs. This information enables you to calculate the timber volume that is available for sale.

The calculations are similar to those described in Chapter 4, for preparing your woodlot inventory. To calculate the board feet per acre, the tree marker will add all the marked trees of each species and calculate the average diameter. Using the heights of the trunks, he will determine how many 8- or 16-foot (2.5 or 5 m) logs can be cut out of each tree. Lengths smaller than 8 feet (2.5 m) are not acceptable sawlogs.

The board feet can be calculated from a table such as the Simplified Board-Foot Volume Table (Table 15) provided below. There are several different variations of these tables in use in different regions, for slightly different purposes. You can also calculate timber in cubic meters, but sawmills still largely operate with standard 8- or 16-foot (2.5 or 5 m) logs and measure them in board feet.

Table 15 Simplified board foot volume table

	MERCHANTABLE LOG LENGTH (FEET)			
DIAMETER	8	16	24	32
10	16	32	44	53
12	25	50	67	82
14	36	73	99	122
16	49	98	135	166
18	62	125	173	215
20	78	157	218	272
22	96	193	269	338
24	117	234	328	401
26	137	275	387	485
28	161	323	456	573
30	187	374	529	667

Source: S.N. Staley. Wood: *Take a Stand and Make it Better.* Toronto: Ministry of Natural Resources, 1991.
G.R. Goff, J.P. Lassoie, and K.M. Layer. *Timber Management for Small Woodlands.* Ithaca, NY: Cornell Cooperative Extension Service. 1994.

The price you receive for your timber will be given as a price per thousand board feet. Prices can vary enormously according to location and demand; for example, in late 1997, sugar-maple sawlog prices ranged from $80.00 to $1,000.00 (Canadian) per thousand board feet, just within Ontario. Prices also vary according to quality.

Logs are graded when cut, with the highest-quality veneer logs often bringing a substantial price bonus; but again, this is dependent on local markets.

The Logging Prescription and Contract

Just as a doctor will write a prescription for a patient, a forester can write a prescription for the logging of your woods. This serves to document the timber to be cut and the environmentally sensitive features to be protected. This document will be very brief, but will pull together the information from the timber inventory, documenting how much is to be cut and any other specific features such as den trees, streams, springs, or wildlife habitats that should remain undisturbed. You should work with the forester to agree on the prescription, which will, in turn, provide guidance for the tree marker and logger.

The impact of a harvest on your woodland will depend heavily on how the logging is carried out. Read Chapter 5 carefully to ensure that you know what features to protect, so any logging has a minimum environmental impact. As well, pay particular attention to the "Woodland Code of Ethics" below.

When landowners are upset at the devastation caused by logging practices, it is usually because the operation was not carried out carefully enough. Careless road construction, using heavy equipment in wet weather, damage to residual trees, and leaving uncut slash lying in the woodlot are the most common causes of concern. As the landowner you can include clauses in your logging contract that prevent such problems, although you may have to do some negotiating to get the logger to accept them.

A local professional (independent of the logging contractor) can be your best friend when it comes time to sell your timber. He can advise you on the best timber markets and assist you in getting bids on the harvest. You should always get two or three bids, and not accept the first one offered. Above all, do not deal with a logging operator who offers you an extra payment in return for taking "a few extra trees".

A contract for the sale of your timber should include the following:

- price and payment method; payment just before harvesting is strongly recommended
- which trees are to be cut and how they are marked
- responsibility for cleanup, including handling of slash and repair of roads
- a schedule or time limit
- liability insurance and an arbitration method
- a penalty for cutting unmarked trees or damaging the trees left in the woodlot
- other clauses as you wish to reflect the code of ethics below

Government agencies or woodlot associations can provide standard logging contracts that you can use or adapt to your own situation.

The Landowner's Role in Logging

It is the landowner's responsibility to arrange with logging companies to market and harvest his woodland. Professional foresters can also help you with this task if you wish, and at the same time provide valuable inspection services during harvesting operations.

The critical things to remember are to

- base your harvest operation on a good woodland inventory, including environmentally sensitive areas to be protected,
- have a professional evaluation of harvest plans to ensure future woodland stocking levels,
- have a written contract with loggers to ensure appropriate harvest practices and cleanup,

- consult and follow all local tree-cutting bylaws or other regulations, and
- inspect your woodlot during harvesting operations, or hire an independent forester to do so for you. You only have to appear in the woods occasionally to ensure that loggers will be much more careful.

Professional foresters can also assist you in marketing your woodland through a timber sale, to ensure you receive an optimal return. *Do not agree to sell your timber if approached by a timber buyer without first seeking independent professional advice.*

Firewood Production

Can you cut firewood without damaging your woodlot? Yes, but it all depends on how intensively and how carefully you plan the cut.

As with timber harvesting, you should not cut or haul out firewood in wet conditions, or in the spring and summer nesting season. See the code of ethics on page 137 for basic guidelines.

It's easy to make three common mistakes in cutting firewood:

1. You may be tempted to go through your woodland and cut dead or dying trees for firewood. Don't. Dead trees are snags, vital habitat and food sources for many species of birds. Furthermore, once they are dead, they are not competing with other trees for growth. Large dead trees are particularly valuable as den trees for wildlife. Trees that provide food for wildlife, such as acorns or beechnuts, are also worth protecting; such trees are called mast trees, and should be conserved.

2. You might look for big old trees that you can cut and then split into firewood. Be careful. Some large trees might make good-quality sawlogs. Cutting them into firewood is like burning money. They might be worth far more as timber.

3. You could identify species that are particularly good
   for heat production, such as beech and hop hornbeam
   or ironwood, and cut only these species. Don't. If you
   do, over time you will shift the balance of species in
   your woods, losing biodiversity and losing these
   species as future firewood. Always cut trees to maintain the balance of species that exist naturally in your
   woodland.

There are two proper ways to cut firewood in your woodland.
First, during any timber harvest, the tops of trees can be cut up for
firewood. Some branches can be cut to make brush piles; this will
provide wildlife habitat. Others will make good firewood. No treetops should be left uncut after a timber harvest.

Secondly, a thinning cut can be planned to improve the growth
potential of crop trees, while producing a harvest of firewood in the
short term. Even if you are only cutting a little firewood, you can
choose the trees to cut as if you were thinning your woodland. In

this way, you do no damage to forest growth and may actually enhance it while getting the firewood you need.

Refer to the discussion of thinning and harvest cuts in uneven-aged and even-aged stands in the previous pages. Plan the trees you will cut for firewood as if you were thinning your woodlot for improving the growth of the remaining trees. Cut poorer-quality or crowded trees while protecting crop trees, and keep in mind the spacing rules described above; the illustration on page 120 provides a visual impression of this management.

Once your woodland is a good-quality, uneven-aged stand, every harvest will generate firewood as well as sawlogs. Under proper management, you should be able to harvest half a cord per acre every year from a good-quality hardwood woodland (see page 91 in Chapter 4 for a discussion of measuring firewood).

Economic Benefits

Landowners are frequently reluctant to provide details of the economic return from a timber harvest in their woodlot; however, anecdotal evidence suggests that a well-managed woodland on high-quality soil can bring substantial financial benefits, in some cases as high as the return from agricultural crops. We have heard reports of economic returns ranging from $80.00 (Canadian) per acre per year to $200.00 per acre per year. Of course, this return only occurs as a lump sum every twelve to fifteen years, and only on well-managed sites.

Any specific statements on financial returns from woodland management depend heavily on local market conditions, site-specific soil, and climate quality, as well as the quality of the existing woodland. We are convinced, though, that good management based on proper understanding will be financially profitable in the long run, provided site conditions are adequate.

Sustainable economic returns depend heavily on resisting the temptation for a quick profit through high-grading your woodland. The return from an individual harvest cut may be slightly lower, but sustainable harvesting will be possible if it is properly done. High-grading may set the harvest potential of your woodland back by several decades, while sustainable harvests should be possible on good soils every twelve to fifteen years.

Growing an Old-Growth Woodland

It is worth noting that the same principles which enable sustainable timber harvesting also enable you to grow large trees faster, or to grow an old-growth woodland, if that is your goal.

Many woodlands have been left in a degraded state through past high-grading, leaving an unnaturally high proportion of diseased, poorly formed, or crowded trees. Properly planned thinning can assist in the transformation of these woodlands into healthier, uneven-aged forests, with appropriate proportions of trees of different quality. In particular, thinning can assist in the faster growth of large trees.

If your interest is in growing old growth, you never carry out the ultimate timber harvest; as your woodland develops in maturity, you protect it for its old-growth characteristics. A professional forester can give you good advice on achieving this goal.

A Woodland Code of Ethics

The major impacts of timber harvesting on biodiversity come through careless or too-intensive logging, or through the removal of woodlands through clear-cutting. With care, timber harvesting and conservation can go hand in hand.

1. Prepare a detailed forest inventory in advance of harvesting, and avoid disturbing special biological features, such as rare species or supercanopy trees, during logging.
2. Avoid disturbing any areas of wet soils, including springs, seepage zones, wetlands, ponds, and streams.

Leave a buffer of at least 100 feet (30 m) around all water features.

3. Plan your timber harvest in order to avoid important wildlife habitats, such as snags (dead trees), large hollow trees that may serve as den trees, and trees that provide food, such as nuts and berries.

4. Minimize road layouts, and keep skidding trails as straight as possible; cut bumper trees along the trails last to protect remaining trees from damage.

5. If your woodland road must cross a stream, install a proper culvert or other crossing with care, and minimize disturbance; if you must access wet areas, do so when the ground is thoroughly frozen.

6. Avoid conducting any timber operations in your woodland during the breeding season for birds, from March until the end of July; this is also the season when bark on younger trees is most susceptible to damage.

7. Have logs winched to skidding trails, or consider horse logging. Skid logs only when ground is frozen or hard and dry; never skid whole trees.

8. Have a professional forester prepare or check your harvesting plan and mark the trees for harvest, taking into account the environmental protection you want.

9. Seek prices from more than one logging company, and get the advice of a professional forester before selling any timber; your net return will likely be higher.

10. Have a clear written contract with loggers to ensure that they follow the above guidelines and minimize disturbance; include a penalty for damage to residual trees.

11. Insist on reasonable cleanup operations, including cutting treetops into firewood or brush piles for wildlife and back-blading trails.

12. Remember that it is your role as the landowner to set the rules for the harvest and to inspect the operations to ensure that loggers comply; otherwise, hire a professional to do this.

13. Set a harvest level that will leave your woodland well stocked for sustainable growth until the next harvest; do not succumb to the temptation of a small extra financial return in the short run while setting back the potential financial return in the long run.

If you are a landowner who is planning to harvest firewood or timber, you should, as a minimum, conduct your harvest operation with care, while respecting the principles outlined in this chapter.

A Word on Safety

Logging is a dangerous activity, requiring specialized training and safety equipment; in some jurisdictions a license may be required to operate a chain saw. Many landowners routinely work in their woodlots, and if you do so, be sure you have the proper safety training and equipment. If you are not careful, chain saws and falling trees can lead to serious accidents.

Proper chain saw–safety equipment includes safety glasses and gloves, heavy steel-toed boots, ear goggles, and a hard hat. Many loggers now wear safety chaps as well. If you want to learn to operate a chain saw safely, training is often available at workshops. A number of books, such as *The Harrowsmith Country Life Guide to Wood Heat,* instruct the reader on how to use a chain saw in tree felling.

Further Reading

Anderson, H.W., and J. Rice. *A Tree-Marking Guide for the Tolerant Hardwoods Working Group in Ontario.* Toronto: Ministry of Natural Resources, 1990.

Anderson, H.W., et al. *A Silvicultural Guide for the Tolerant Hardwoods Working Group in Ontario.* Toronto: Ministry of Natural Resources, 1990.

Fazio, James R. *The Woodland Steward.* Moscow, ID: The Woodland Press, 1985.

Goff, Gary R., James P. Lassoie, and Katherine M. Layer. *Timber Management for Small Woodlands.* Ithaca, NY: Cornell University Extension Service, 1994.

McEvoy, Thomas J. *Introduction to Forest Ecology and Silviculture.* Burlington, VT: University of Vermont, 1995.

Thomas, D. *The Harrowsmith Country Life Guide to Wood Heat.* Charlotte, VT: Camden House, 1992.

Van Ryn, Debbie M., and James P. Lassoie. *Managing Small Woodland for Firewood.* Ithaca, NY: Cornell University Extension Service, 1987.

Walker, Laurence C. *Farming the Small Forest: A Guide for the Landowner.* San Francisco: Miller Freeman, 1988.

CHAPTER 7

REFORESTATION

Whether your interest is in conserving biodiversity and watching wildlife or in firewood and timber production, reforestation to expand your woodland and connect it to other woodland patches is one of the most important management options you can choose. You may choose to reforest gaps between smaller woodland patches, buffers around woodlands, wetlands or waterways, or reforest connecting corridors between woodlands. You may have patches of marginal land that are better reforested on your farm, or if your land is former farmland, old fields may already be growing up in trees.

Reforestation can also include such agroforestry options as planting Christmas trees, planting windbreaks and fencerows, and planting valuable hardwoods along these corridors. These specialized options are dealt with in Chapter 8.

You also have several choices in reforestation. You can plant conifers such as white and red pine, tamarack, cedar, or spruce in a plantation, or you can plant some hardwoods such as ash, oak, or walnut. You can also choose to let natural succession take its course. But if most of the landscape on your property or in your region is already woodland, you may wish to protect the open-field habitat instead of reforesting it; here your choice is among techniques for maintaining the open habitat.

You should also be able to recognize rare and unusual open habitats that should not be reforested. Prairies, wet meadows, savannahs, or rocky habitats, such as "alvars," are unique habitats that make a major contribution to natural biodiversity; they should be protected from reforestation.

In this chapter, we first examine choices for reforestation, including some of the reasons you may choose to plant trees. We then discuss three options for reforestation: natural succession, planting hardwoods, and planting coniferous trees. In the latter case, we also discuss the ongoing management that will be required; evergreen-plantation management is one of the most neglected topics in woodland management. Finally, we review cases where you should not allow reforestation and methods for maintaining open habitats.

Choosing Reforestation

There are several reasons to choose reforestation as a management option. First and most important, current scientific research suggests that the conservation of biodiversity will be enhanced if woodlands are larger, shaped as solid blocks of forest, and connected. Therefore, planting to expand your woodland or create connecting corridors of natural vegetation is recommended. Corridors can simply be existing fencerows; they should be as wide as possible and continuous. If your woodland has been broken into small patches through past clearing, replanting to create a more solid block of forest is recommended.

Planting or maintaining a buffer of natural vegetation along streams and around wetlands is also a highly recommended practice, both to maintain water quality and for wildlife habitats. You can combine this with the creation of natural corridors. Planting can also add variety to natural habitats on your land. If your wood-

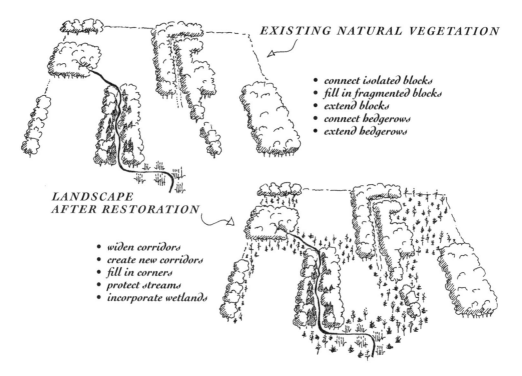

Expanding existing woodland to create larger solid blocks of forest and connecting woodlands with vegetation corridors are the two most important contributions to conserving biodiversity in any landscape where woodland is fragmented.

land is entirely deciduous, planting an area of conifers will add variety; if you have a coniferous plantation, interplanting or seeding with deciduous species can be part of a management plan, as discussed below.

Finally, if you have marginal land on your farm, it may be wise from both environmental and economic points of view to consider reforesting it. Marginal land usually consists of smaller patches of land on a farm that do not produce a marketable crop because they are too wet, too steep, too rocky, or the soil is too erodable. In fact, many farmers have incorporated such patches into field production simply for convenience; it is easier to clear or ignore such obstacles and plow the entire field rather than to work around them.

However, recent research in precision farming suggests that cultivating such marginal land may actually represent a financial loss for farmers. Through the use of continuous-yield monitoring and tractor-mounted, global-positioning systems, farmers can now produce a detailed map of yield across their fields, instead of just a gross average yield per field. These detailed maps show areas of low yield on patches of marginal land. If the cost of fertilizer and seed exceeds any crop gained on these patches, you are losing money by including them in your cropland. You would make more money by leaving them alone. You could make significantly more money by planting these areas with Christmas trees or hardwoods.

Natural Succession

The easiest approach to reforestation is to allow natural succession to take its course. As described in Chapter 2, a variety of species will gradually invade your land, changing it from an open field to forest. This option provides a sequence of varied habitats and their associated wildlife for you to enjoy over time.

One specific recommendation we suggest is to begin mowing a walking trail through your field when it is still at an early successional stage, before any woody vegetation appears. If you start in the early spring, before the grass gets too high and dense, you can do it with an ordinary garden lawnmower (a riding mower is easier). Otherwise, your old field will soon become difficult to walk through and, therefore, difficult to enjoy, as succession progresses. Such trails also make for great cross-country skiing in the winter. Maintained long enough, they will eventually become the basis for woodland trails. See Chapter 9 for suggestions on trail routings.

Several variables will determine how quickly an old field grows up in shrubs and trees. Remember that all plants are competing for soil nutrients, sun, and moisture as they grow. But species such as grasses that live in the open field are usually more efficient in this competition than trees, especially deciduous trees. A dense, healthy stand of grass or weed cover in an open field can maintain itself for up to several decades before much succession to other

If a good seed source is available and competition from grasses and weeds can be controlled, natural succession may be the fastest way to recreate a woodland.

species occurs, especially if there is no alternate seed source nearby. If this is the case on your land, the best option may be to leave it alone and enjoy it. As stated previously, in some parts of eastern North America, open-meadow habitats and their wildlife species are increasingly rare.

If a farm field with a healthy stand of grass is abandoned, natural succession may be very slow. On the other hand, if a field is plowed before natural succession begins, the mineral soil is exposed with all its dormant seeds. This provides for a more rapid start in the invasion of other species. Available seed sources are also a major influence. If there are seed-producing trees in a fencerow downwind from an open field, tree seedlings may establish themselves very rapidly. Seed sources and bare mineral soil provide the basis for the most rapid tree establishment.

The species of trees used as a seed source make an important difference. Trees that enjoy open sun (trees that are intolerant of shade), such as trembling aspen, white birch, and white ash, may spread very rapidly in these conditions. Sugar-maple seedlings may germinate and start growing, but will grow only slowly in open-sun conditions, often suffering from sunscald. Some trees such as trembling aspen will also spread vegetatively, sending new trees up from spreading underground roots.

When plentiful seed sources are nearby, and bare mineral soil can be exposed, natural succession probably provides the fastest way to reestablish a deciduous forest, especially if soil conditions are good for growth. Young ash saplings can create a low partially shaded canopy in only fifteen to twenty years. If grass cover is dense and nearby seed sources are unavailable, some active reforestation can speed up the process of woodland reestablishment, should you choose to do this.

If natural succession can provide reforestation, why bother with plantations of pine or spruce to do it? The reasons are simple. Coniferous plantations were developed to deal with eroded soil where establishing forest cover was a difficult challenge. Often there is a dense stand of grass present; conifers compete with grass and survive much more readily than do hardwoods. Also, there is not always a natural seed source available to speed up regeneration, and coniferous trees provide a crop in a relatively short time, although they do not have the value of a hardwood forest.

It must be admitted that coniferous plantations have become a bit of a habit, especially after government agencies invested funding to provide tree seedlings and equipment. In some cases, however, natural succession may be an equally fast way to establish a woodland.

You can, of course, combine natural succession with active reforestation, either with hardwoods or conifers. Instead of planting an entire area with trees, try planting a few scattered trees of various species or plant patches of trees within the field, allowing natural succession to do the rest. Read the following two sections for advice on tree planting.

If you wish to establish a forest as fast as possible, a full-scale plantation of coniferous trees will often be the best answer, as long as you manage it correctly after planting. For small areas, plantations of deciduous trees may be used; they need extensive tender-loving care in the first few years to get established. The following section provides a detailed description of tree planting, followed by a discussion of the advantages of different approaches.

Tree Planting

Tree planting begins with an inventory as described in Chapter 2. For reforestation, an inventory should pay particular attention to soil and moisture conditions. Trees must be selected to match the conditions on the land you are going to plant. The inventory should be done in the spring, if possible, when you can recognize the wet patches that may be dry by August.

Table 16   Soil conditions for tree species

SOIL TEXTURE	SOIL DRAINAGE	
	GOOD TO MODERATE	IMPERFECT TO POOR
Sand	White pine Red pine Sugar maple Red oak White cedar Poplar	White cedar Tamarack Black spruce Green ash Willow
Loam	White pine Red pine White spruce Sugar maple Red maple White cedar White ash Green ash Red oak Black cherry Beech Basswood Black walnut Poplar	White cedar Tamarack Black spruce Silver maple Red maple Willow
Clay	White pine White cedar White ash Green ash Beech Red maple Black walnut Poplar	Tamarack Black spruce Silver maple Green ash Willow

Source: Landowner Resource Centre, 1995. *Planning for Tree Planting (fact sheet)*. Manotick, ON: Landowner Resource Centre.

Secondly, you need to develop a planting plan. This depends, above all, on your own objectives. You may be planting to retire land or encourage wildlife, to grow Christmas trees, or to eventually harvest timber. Local experience will help tell you which species will do best in your area, but some suggestions are provided in Table 17 below.

Table 17   Trees to select

PURPOSE	SPECIES
Coniferous plantations for timber production	*White pine*
	Red pine
	White spruce
Coniferous plantations for Christmas trees	*Scots pine*
	White spruce
	Balsam fir
Deciduous plantations for timber production	*Black walnut*
	Black cherry
	White ash
	Red and white oak
Plantings where some shade is available	*Sugar maple*
Windbreaks	*White spruce*
	Eastern white cedar
	Poplar
Wildlife cover and food	*See Table 12, page 104*

Always check the species you choose against soil conditions – see Table 16.
Source: Landowner Resource Centre, 1995. *Planning for Tree Planting (fact sheet)*. Manotick, ON: Landowner Resource Centre.

The species you choose should also be based on the size of your planting area and the effort you can put into tending it. Once planted, a coniferous plantation can be left alone for fifteen or twenty years until it needs thinning; deciduous or hardwood trees will need careful tending for the first three to five years. Larger sites are easier to plant with conifers because of the lower cost of tending them after planting.

The species you can plant will also depend on what is locally available for planting stock. Local government agencies can provide this information. You must contact the tree nursery well in advance in order to see which species can be purchased in the numbers you need.

When considering mixtures of trees, try to mimic nature. Plant trees together that you would usually find together; your own existing woodland is probably the best guide to this. It is also a good idea to plant groups of any one species rather than intermingling individual trees, especially if you are mixing deciduous trees and conifers, or planting only conifers. Some species will grow faster than others and start to shade nearby trees; patches of the same species should enable at least some trees in the group to survive to full size. Remember, from the thousands of trees that start out as seedings in every acre of natural woodland, only two or three hundred will make it to maturity.

Finally, you need to decide on the spacing of trees to be planted; this will help you know how many trees to order. Spacing depends on the ultimate purpose of your planting; alternatives are outlined in the table below.

Table 18   Spacing for tree planting

REFORESTATION TYPE	SPACE BETWEEN ROWS	SPACE BETWEEN TREES
*Coniferous plantations**	*8 ft. (2.4 m.)*	*6 ft. (1.8 m.)*
*Deciduous plantations**	*10 ft. (3 m.)*	*5 ft. (1.5 m.)*
Maple syrup orchard	*30 ft. (10 m.)*	*30 ft. (10 m.)*
Christmas trees	*5 ft. (1.5 m.)*	*5 ft. (1.5 m.)*

* A coniferous plantation planted at this density will have 900 trees per acre (2300 trees per hectare); a deciduous plantation will have 880 trees per acre (2000 trees per hectare).
Source: Landowner Resource Centre, 1995. *Planning for Tree Planting (fact sheet)*. Manotick, ON: Landowner Resource Centre.

The Planting Plan

Putting your planting information together will provide a plan that shows

- the location of areas to be planted on your property,
- the species to be planted,
- the numbers required,

- the methods you will use for site preparation, and
- how you will tend the trees after planting to control competing vegetation

Do not expect to plant trees everywhere; wet sites and shallow soils over bedrock are probably best left unplanted. Look for unique open habitats as well; these are discussed later in this chapter. You should also plan your layout of trails or roads at this time, and leave them unplanted; this is far easier that clearing trails later.

Some landowners forget to think of the views on their property when planning reforestation. Especially near your home, it is worth considering views that you would like to maintain; designing these areas into your planting scheme as "viewsheds" to be left open will later enhance the aesthetics of your land. Though it may appear very open when you start planting, you can feel very closed in ten or fifteen years later when trees are growing vigorously.

With your planting plan prepared, you are ready to prepare the site for planting and to order trees. It is worth the trouble to share your reforestation plan with a professional to ensure that your choices are appropriate for your area.

Site Preparation

Site preparation refers to the removal of competing vegetation to provide bare soil for planting trees. If existing vegetation is fairly light, clearing a strip 4 feet (1.2 m) wide, or a circle 4 feet (1.2 m) across, for each tree will be sufficient; this is usually done with spot spraying or band spraying of a herbicide. It can also be done by light cultivation. On a small scale, you can do this with a rotary cultivator; on a large scale, a tractor and plow will be required. It may be necessary to mow down existing vegetation before cultivating.

Herbicides must be applied with care, and their application requires special equipment, such as a backpack sprayer or a tractor-mounted boom sprayer. Some jurisdictions require you to take safety training and be certified before you can purchase herbicides. Consult a government agency for advice on the procedures in your jurisdiction and for advice on current products.

When vegetation is quite heavy or dense, site preparation may be more difficult. It could involve the mechanical removal of brush, full cultivation, or herbicide applications. Many common species of shrubs will resprout quickly when cut down; this can be controlled by repeated cultivation or by allowing a period of regrowth and then applying herbicides. You may be better off to avoid areas of dense brush growth.

Wet sites are particularly difficult to plant trees in, since trees will not survive well with wet roots. It may be better to leave these sites alone as well. Otherwise, furrowing can be used to build up a ridge of drier soil, then you can plant trees on top of the over-turned furrow.

Schedules

The best time for tree planting is in the spring. Since this does not usually allow time for site preparation first, this should be done the previous fall. This implies that a tree-planting project has at least a twelve-month schedule, from the time you begin your inventory in the spring, through the development of your planting plan, to site preparation, and the ordering of tree seedlings in the fall. Trees will be planted the following spring, during the brief window of time when they are dormant. Some tree species can be planted in the fall, and container-grown stock can be planted at other times, but check with your nursery supplier to be sure.

The schedule on the day of planting is also important, because trees should be planted immediately upon arrival, if possible. Inexperienced tree planters may be able to plant 300 to 400 trees a day; experienced planters may plant more than double this. If your site is dry and relatively stone free, you may be able to use a tree-planting machine and plant several thousand trees per day. With a spacing pattern that requires about 900 trees per acre (or 2,000 per hectare), you may want to plan your tree planting in stages over several years, unless you can hire an agency or contractor to plant the entire area for you at once.

Planting Your Trees

Tree seedlings of all kinds need extra tender-loving care before planting. From the moment they are shipped to you they are under stress. Transport seedlings so they do not overheat, and minimize travel time; treat them as if they were cartons of eggs, not bales of hay. Unload seedlings immediately; store them in a cool, shaded spot with stream water nearby to water them with, if possible. Bare root stock should be planted the same day; all seedlings should be planted within thirty-six hours.

When planting your tree seedlings, the goal is to get them out of the cool storage bag and into the ground as quickly and as carefully as possible, without letting the roots be exposed to the air or dry out. Think of seedling roots as having an allergic reaction to sunlight. Hide them from the sun as much as possible; even a few moments in a drying breeze or direct sun can stress them beyond survival. Place a little cool water in the bottom of your planting pail, and put seedlings in the pail quickly; separate them carefully, lifting only one seedling out at a time when you are ready to plant it. Keep all storage bags closed.

Finally, plant each seedling with care. Choose a clean spot where you can dig a deep enough hole. First, slice the shovel vertically into the soil, then on an angle to the first cut to lift out a wedge of soil. Place the seedling carefully into the hole. Replace the wedge of soil and tamp the dirt back in the hole with your toe. Do not rush to plant your trees; a little care now makes a major difference in their survival and growth.

Pay attention to the details of carefully handling and planting seedlings; survival rates can be quite low when untrained volunteers are used for tree planting, especially with bare-root seedlings.

Careful tree planting is essential to seedling survival.

Container or Bare-Root Stock?

Container-grown stock varies from small seedlings grown in a small plug of soil to large stock that comes in a pot or with a burlap ball of soil around the roots. Container stock, especially in large sizes, is always more expensive than bare-root stock, sometimes much more expensive. The advantage of container stock is that roots will undergo less shock during planting, since soil is not removed; they can also be planted more readily at different times of the year. Handle container stock by the plug or burlap ball, not the stem.

Bare-root seedlings must be planted quickly to ensure that their roots are not exposed to air for long, and they can only be planted during brief periods in spring or fall when the plants are dormant. However, they are usually much cheaper.

For large-scale reforestation, relatively small bare-root seedlings are usually used, because of the number of trees required and the cost. For smaller-scale projects and urban projects involving community groups, larger stock grown in containers is often used. In the authors' opinion, planting stock much larger than 3 feet (1 m) high is not worth the cost outside an urban or backyard setting. The transplant shock sets larger trees back enough that smaller trees catch up easily over a year or two. Besides, part of the fun is in watching them grow.

Planting Seeds or Moving Your Own Trees

Some nurseries can provide tree seed for planting. It is also possible to gather seed yourself from some trees such as walnut or oak. You can then establish your own small-tree nursery and grow trees to transplant size, or else plant seed directly. Most tree seeds require some period of time at different temperatures, similar to what they would encounter in the outdoors, in order to germinate. Consult a basic text that includes notes on propagation in order to determine what is needed for any one species. The recently published *Trees of the Central Hardwood Forests of North America* provides this information. Any nursery that sells seed will also be able to tell you how to plant it.

A small-tree nursery will require fencing to keep out moles, mice, and rabbits, and also requires shading, which is often provided by placing old snow-fence slats over the planting bed.

You can also transplant trees from your own land, if you have them available. It is best to find trees that are already growing in an open habitat, such as along fencerows; trees moved from the shade of a woodlot to bright sun may not thrive. They may also be quite old trees, their growth stalled in the low light beneath the canopy. Look for saplings that are putting on a foot or more of growth per year as your transplants. This provides trees at minimal cost, and you can be sure they are adapted to your locality, although transplanting them will be a lot of work.

After the Trees Are Planted

Young conifers can usually be left to compete with grasses and other plants in the field; they will gradually rise above this competing vegetation and begin to grow more rapidly. However, deciduous trees need to be tended carefully in order to compete successfully for moisture and nutrients with other vegetation. Tending means removing competing vegetation from a 2-foot (60 cm) circle around the tree, at least until it grows tall enough to rise above the competing vegetation. Since the tending of deciduous and coniferous trees is so different, they are discussed separately below.

Caring for Hardwood Plantations

Reforestation by planting deciduous, or hardwood, trees is a challenging, but rewarding, alternative or addition to coniferous plantations and to natural succession. As noted previously, seedlings of many deciduous trees cannot compete well with grasses and other cover in open-field situations. Deciduous trees have greater requirements for both nutrients and moisture than do conifers. When planted for reforestation, deciduous trees will, therefore, need extended protection in order to survive. This is the primary reason that conifers are so often used for reforestation instead.

Tending deciduous trees can be done manually, by hand weeding or cultivating, or through the use of mulch, if you are dealing with only a small area. The most common mulch is wood chips, but you can also use newspapers or old carpet, and you can purchase squares of fabric or plastic designed for this purpose.

You also require some form of tree guard to protect young trees against voles that eat the tender bark in the early spring and

Hardwood trees planted during reforestation require protection with tree guards and mulch.

against mechanical damage from weed whippers. You can use either plastic commercial tree guards, hardware cloth (a ½-inch [1.3 cm] wire mesh), or lengths of plastic field-drainage pipe cut lengthwise. Alternatively, you can place tree shelters, which are plastic tubes about 3 feet (90 cm) high, over the trees. This is expensive, but it also contributes to faster tree growth and a straighter trunk. Regardless of which tree guards you choose, you still must mulch, for the competition that counts is that going on underground among the roots of trees and grasses.

In larger deciduous plantings, tending must be done using chemical spraying. Two primary types of spray are available. A pre-emergent herbicide, such as symazine, can be sprayed right over the trees and will kill weed seeds before they grow. A post-emergent herbicide, such as glyphosate, will kill weeds after they have

sprouted. A pesticide-applicator's license may be required to purchase these chemicals; your local dealer can advise you.

Since this spray will kill all vegetation, trees must be protected.
A common recommendation is to buy a short length of stovepipe,
put a handle on it, and place it over individual young trees while
spraying around them.

With a backpack sprayer, you can cover a lot of trees fairly
quickly, but herbicides are expensive, and their use should be mini-
mized to prevent any harmful impact. Chemical tending can, therefore, be combined with mowing the strips between the rows, or
with planting cover crops, such as white clover or a rye grass. This
minimizes the spraying to just the immediate area around the tree.

For a large area, this level of tending represents a substantial
investment of energy, time, and cost. It is easy to see why coniferous plantations are the popular alternative, requiring little such
attention.

For small areas, deciduous trees such as oak, walnut, or ash can
be planted as the primary form of reforestation, or they can be
planted in clumps mixed with conifers. They can also be planted in
old fields that are undergoing natural succession or along
fencerows and windbreaks. One farmer we know has several thousand black-walnut trees in fencerows around his farm. At a value of
up to several thousand dollars per sawlog, these trees will make his
grandchildren wealthy indeed.

Caring for Coniferous Plantations

Coniferous trees such as pine, spruce, and cedar are the most common trees used for reforestation. Usually planted in solid plantations, these trees tend to survive well when initially planted and
eventually form a shade canopy that will allow the reestablishment
of hardwood-forest species. Conifers are generally intolerant trees
(except hemlock), growing in sun rather than shade. Some coniferous species, especially red and white pine and white spruce, can be
managed for timber harvesting, providing sawlogs when they are

In many cases reforestation with conifers is the best first step in recreating natural woodland, with thinning required over time.

mature. Alternatively, coniferous plantations can be managed to allow for a gradual transition to deciduous forest, as described below. Unfortunately, many such plantations are not managed at all, creating neither useful timber nor a more diverse forest.

Initially, trees are usually planted in rows about 8 feet (2.5 m) apart, with trees spaced about 6 feet (1.8 m) apart. Although some complain that these rows look artificial, they are, in fact, the fastest way to grow a crop of trees. Since you are dealing with 900 trees per acre (2,300 trees per hectare) planted at this spacing, machine-assisted planting is much more efficient, unless you have only a very small area to plant.

The primary advantage of planting conifers to start your reforestation is that these species have relatively low nutrient and moisture requirements. They can survive the competition of nearby grasses and weeds, which are common in old-field situations. After three to five years, they will poke their branches above this competing vegetation and take off for the sky, growing very rapidly if conditions are good. It is common for these species to put on 2 to 3 feet (60–90 cm) of growth per year once they pass about 8 feet (2.5 m) in height. After they are planted, little or no tending is required for many years, though early tending will certainly assist initial growth.

As noted previously, coniferous plantations can be enhanced by also planting shrubs to provide food and cover for wildlife and by planting some hardwood trees, either during initial establishment or when thinning. Putting some thought into the area to be planted and keeping trails and viewsheds open is particularly important with solid coniferous plantations.

Plantation Management

The original theory behind planting solid coniferous plantations was that they would control soil erosion and create a shade canopy quickly, thus enabling the regrowth of hardwood forest. However, this does not happen all by itself. Proper management of the conifers is required through thinning and seeding in of hardwood species, either naturally or with your assistance. Unfortunately, many landowners are either unaware of this or unable to do the work, and many government agencies have found that budget cuts prevent proper plantation management.

Thinning should begin when trees are between fifteen and twenty-five years of age, when they are approaching 30 to 40 feet (9 to 12 m) tall and looking crowded. If planted in rows, every fourth row can be removed at this stage. Unfortunately, there is little market for these first thinnings, though they can be used for small poles, lumber, firewood, or if a local market is available, for pulpwood or chips.

After the first thinning, plantations should usually be thinned every ten to twenty years to maintain optimum growth. The goal is to allow the growth of large, high-quality poles or sawlogs, while keeping enough closed canopy so that lower branches will die over time. Red and white pine are described by foresters as "self-pruning"; their lower branches die naturally as long as they are growing in a crowded plantation environment. However, you can also prune trees at this stage to ensure straight sawlogs with few knots.

Thinning coniferous plantations is essential to encourage a transition to hardwood forest.

The second thinning ten to fifteen years later (or sooner on good sites) can remove alternate rows, leaving only half the original trees. After this, further thinnings can concentrate on removing smaller trees of poorer form, leaving the best crop trees. White and red pine can be managed to produce merchantable sawlogs or utility poles in eighty to 100 years, though some of the tallest pines should be left as future seed sources and as supercanopy trees for perching birds. By the third thinning, harvested trees should provide good-quality utility poles and sawlogs.

The benefits of planting in rows become apparent when thinning, for thinning is relatively easy in rows, but becomes very difficult otherwise. Eventually, when entire rows and individual trees within rows have been removed, the woods will begin to appear more natural.

Deciduous species such as maple, ash, and oak will seed in naturally if thinning is carried out and a natural seed source exists nearby. This is the process you are hoping for when you plant a coniferous plantation. But this can be supplemented by either direct planting or by spreading seed. It is often useful to plant seeds in small gaps or openings you have cut in the plantation while thinning. Plan machine access to your plantation carefully, in order to avoid planting seedlings in areas later needed as access roads for future thinning or harvesting.

Cutting small openings or gaps will also speed up the natural regeneration of hardwoods within coniferous plantations. When you are thinning, cut two or three small openings to allow the hardwood seedlings to establish themselves. Smaller gaps under 35 feet (10 m) across will favor shade-tolerant species, such as sugar maple, beech, and hemlock; larger gaps up to 165 feet (50 m) across will encourage less-tolerant species, such as white ash, oak, and perhaps black cherry, if seeds are available.

With proper management, coniferous plantations can be a practical option for restoring natural woodlands. Once well established, they require thinning at regular intervals, but they are relatively easy to establish in the first place. With enough time and care, a natural hardwood stand will be reestablished. With proper thinning, a plantation can begin to look like a mixed forest within thirty to forty years.

Planting for Wildlife

At the same time as you are planting trees for reforestation, you can enrich the habitats you are creating by planting shrubs, which will provide either food or cover for wildlife. Table 12 in Chapter 5 provides a list of species to use.

Shrubs should be planted in clumps, which will form patches of cover for wildlife. As with trees, they can also be planted in old fields that are undergoing natural succession, during the original planting of trees for reforestation, or during thinning of conifer plantations.

Avoiding Invasive Exotic Species

It is also important to avoid the use of invasive exotic species. Although in the past some nonnative tree or shrub species, such as Norway maple, black locust, Norway spruce, Russian olive, and others, have been recommended for reforestation projects or for wildlife plantings, today the use of nonnative or exotic species is not encouraged. In too many cases, such species have turned out to have had negative ecological impacts.

For example, Scots pine tends to develop a crooked trunk, giving it no timber value, and spreads easily into old fields once it is mature enough to produce cones. Norway maple has heavy, thick leaves that smother native wildflowers on the forest floor, gradually destroying the woodland groundcover; it is also prolific in producing seeds, which germinate more readily than the native sugar maple.

Table 19 provides a list of invasive exotic species you should avoid and native species to use as substitutes.

Using Genetically Appropriate Planting Stock

In addition to avoiding exotic species, you should always use planting stock that is genetically suited for your region. Ideally it should be grown locally, from local seed sources. Locally, in this case, refers to the same basic climatic zone and the same physical landscape.

Populations of trees contain substantial genetic diversity within them. For example, sugar maples can be found across a wide range of eastern North America, but within this total population are trees that are somewhat adapted to warmer temperatures or to colder winter conditions, trees that tolerate wetter soils and trees

Table 19 *Avoiding invasive exotic species*

INVASIVE SPECIES TO AVOID	ALTERNATIVES TO USE
Black locust	American highbush cranberry
European alder	Basswood
European birch	Honey locust
European highbush cranberry	Mountain ash
European mountain ash	Pin cherry
Horse chestnut	Serviceberry
Glossy and common buckthorn	Sugar maple
Multiflora rose	Trembling aspen
Norway maple	Trumpet creeper
Russian and autumn olive	Virginia creeper
Scots pine	White ash
Tree-of-heaven	White pine
White mulberry	
White poplar	

Source: J.M. Daigle, and D. Havinga. *Restoring Nature's Place.* Schomberg, ON: Ecological Outlook Consulting, 1996.

that do best on dry sites, and so on. You cannot purchase tree seedlings from distant sources and expect them to do well locally. You also run a risk of diminishing the natural genetic diversity within the population of any one species by importing trees from a different region. Purchase native species of planting stock grown as near as possible to your home.

Buying Trees

In the past, many government nurseries provided planting stock of tree seedlings to rural landowners for almost no charge. Some agencies may still have programs that enable you to acquire trees at very low prices or for free. However, many government tree nurseries are being phased out due to budget cuts. It is, therefore, necessary in these cases to turn to the private sector to purchase tree seedlings. Many private tree nurseries are springing up to feed the demand.

You can ascertain the situation in your own region by contacting your provincial or state forest agency or woodlot-owners' association; contact addresses are provided at the end of the book.

Recognizing Unique Open Habitats

Just as there are numerous opportunities for reforestation, there are cases where you should recognize unique open habitats and protect them. This may require that you call in an expert for assistance, but the knowledge you gain will be well worth it. You may also want to protect some open-field habitats for variety on your land.

Among the unique open habitats that you should look for on your land, and protect if you discover them, are the following:

- Prairie remnants: extending as far east as the east end of Lake Ontario, small patches of prairie vegetation may have tall grasses or unusual species such as the purple coneflower
- Oak savannahs: open, park-like grassland, dotted with scattered oak trees, these are one of our rarest remaining habitats in eastern North America
- Alvars: open, rocky habitats with a unique assemblage of plant species, which depends on the type of bedrock
- Open meadows: many species of wildlife can be found in open meadows, wet and dry; these may only be a stage in natural succession, however, and may require management to be maintained
- Unique wetlands: include bogs, dominated by sphagnum moss, or fens, which look like wet meadows dominated by grasses and sedges
- Sand dunes: because they are often associated with beautiful beaches, most remaining sand dunes are already protected in state or provincial parks
- Rock outcrops or cliffs: often habitat to rare species that have adapted to grow in the crevices of the rock

If you have any of these unusual habitats on your land, you should consult a professional for advice on their conservation. You should be able to find names of experts who may be able to help through a state or provincial natural-resources agency, a natural-history society, a local university biology department, or through the state office of The Nature Conservancy (in the U.S.) or the Nature Conservancy of Canada.

If you find such habitats on your land, the recommended management choice is to protect them, not to reforest the area.

Your choice to reforest an area should also be based, in part, on the landscape in which you live. If your land lies within an intensive agricultural landscape, where remaining woodlands are small and scattered, reforestation to expand woodland cover is probably the best option, unless you have one of the rare habitats described above. If your land lies within areas where forest cover is plentiful and increasing, as in much of the northeast, maintaining open-field habitats will probably be the more appropriate and interesting option. In some regions of the northeast, open fields will be the rare habitat of the future rather than woodland.

Maintaining Open Habitats

There are four practical methods of maintaining open-field habitats.

1. Mowing: The easiest approach to keeping natural succession from taking over is to mow your field once or twice a year. This will probably require a tractor-mounted mower or bush hog and should be done in the late summer or early fall. This will avoid disturbing nesting wildlife species earlier in the season and aid in seed dispersal. If you do not have the equipment, you may be able to hire a neighboring farmer to do this for you for a low cost.

2. Grazing: This is much more difficult to arrange, unless you raise livestock yourself. It requires fences kept in

good repair and, if you do not have your own livestock, a cooperative farmer with an appropriate herd to graze the land for a short time. Grazing in mid to late summer is best, in order to protect nesting wildlife, and should only be done for a long enough time (perhaps a few weeks) to control plant growth. Excessive grazing will result in the survival of fewer species, leaving a degraded natural habitat. You may be able to arrange this at no cost in return for free pasturing, assuming your fences are in place.

3. Burning: For specialized habitats such as prairies, occasional burning is necessary for maintenance. This operation is obviously dangerous and requires professional evaluation. It must be timed for a critical period in early spring or late fall and done under a biologist's supervision. Of course, it also requires fire-department permission and supervision. If you believe that you have a prairie habitat that might benefit from this treatment, consult the experts.

4. Farming: Although it does not create a natural habitat, another option to protect open fields from natural succession is to rent your fields to a farmer for pasture or for hay production. This option at least delays your management decision, protecting your options for the future. Once natural succession has progressed to woody vegetation, restoring an old field to agriculture is considerably more expensive. If pasturing and haying is delayed into early July and if pasturing is kept at a moderate intensity, even these fields provide useful nesting habitats for some bird species.

Establishing Wildflower Meadows

On a very small scale, it is possible to manage open areas by establishing wildflower meadows. This is particularly useful for rural homeowners with large yards or for protecting specific viewsheds from your home. Of course, natural succession will create a wildflower meadow for a period of time as well, but in the immediate home area, an artifical wildflower meadow may be more attractive.

It is first necessary to remove competing vegetation, including grasses or weeds. You can do this by spraying with a wide-spectrum herbicide such as Round-up (be sure to follow directions exactly, and use care with any chemical), by tilling the area repeatedly over one growing season, or by tilling then covering the area with clear plastic for a growing season. Then seed a wildflower mixture into the soil, raking the soil over it to cover the seeds slightly. With native species it is not necessary to water them.

A critical challenge is to find an appropriate seed mixture; many contain some exotic nonnative species or grasses to provide filler. Ask for a seed mixture that has been specifically prepared for your region and contains only native species. Some native grasses are useful components, as long as they are an appropriate species; prairie grasses which form bunches or clumps are often used. Consult a good reference, such as *The Ontario Naturalized Garden* by Lorraine Johnson or *The Wildflower Meadow Book: A Gardener's Guide* by Laura Martin.

Further Reading

Daigle, J.M., and D. Havinga. *Restoring Nature's Place.* Schomberg, ON: Ecological Outlook Consulting, 1996.

Henderson, C.L. *Landscaping for Wildlife.* Minneapolis: Minnesota Department of Natural Resources, 1987.

Johnson, Lorraine. *The Ontario Naturalized Garden.* Toronto: Whitecap, 1995.

Landowner Resource Centre. "Careful Handling and Planting of Nursery Stock" (fact sheet). Manotick, ON: Landowner Resource Centre, 1995.

———. "Clearing the Way: Preparing the Site for Tree Planting" (fact sheet). Manotick, ON: Landowner Resource Centre, 1995.

———. "Cover Crops Help Tree Seedlings Beat Weed Competition" (fact sheet). Manotick, ON: Landowner Resource Centre, 1994.

———. "Managing Regeneration of Conifer Plantations to Restore a Mixed Hardwood Forest" (fact sheet). Manotick, ON: Landowner Resource Centre, 1996.

———. "Mulches Help Trees Beat Weed Competition" (fact sheet). Manotick, ON: Landowner Resource Centre, 1995.

———. "Planning for Tree Planting" (fact sheet). Manotick, ON: Landowner Resource Centre, 1995.

———. "Protecting Trees from Vole Damage" (fact sheet). Manotick, ON: Landowner Resource Centre, 1995.

———. "Room to Grow: Controlling the Competition" (fact sheet). Manotick, ON: Landowner Resource Centre, 1995.

———. "Tree Guards Protect Your Trees" (fact sheet). Manotick, ON: Landowner Resource Centre, 1997.

———. "Tree Shelters Help Hardwood Trees Grow Faster" (fact sheet). Manotick, ON: Landowner Resource Centre, 1995.

———. "Using a Backpack Herbicide Sprayer to Control Weeds" (fact sheet). Manotick, ON: Landowner Resource Centre, 1997.

Martin, L.C. *The Wildflower Meadow Book: A Gardener's Guide*. Charlotte, NC: Faast and McMillan, 1986.

Ministry of Natural Resources. *Managing Red Pine Plantations*. Toronto: Ontario Ministry of Natural Resources, 1986.

TreePeople. *The Simple Act of Planting a Tree*. Los Angeles: J.P. Tarcher, 1990.

Weiner, Michael A. *Plant a Tree*. New York: Wiley, 1992.

Yepsen, Roger B. *Trees for the Yard, Orchard, and Woodlot*. Emmaus, PA: Rodale, 1976.

CHAPTER 8

SPECIALIZED AGROFORESTRY OPTIONS

T here are several specialized management options either asso-ciated with reforestation or with the use of existing wood-lands. "Agroforestry" options are associated with agricultural operations and provide additional uses for trees. These options include the planting of windbreaks and vegetation buffers, the planting of hardwood-crop trees, Christmas-tree growing, and the production of maple syrup.

Each or all of these can easily have their place on rural property, and range from very small-scale, personal hobbies, to full-size, commercial-scale operations. In this chapter, we review the details of each of these four agroforestry options.

Windbreaks and Vegetation Buffers

Planting windbreaks to protect farm fields from erosion, to provide an energy-saving barrier around your home, or to provide connect-ing wildlife corridors between woodlots is just a specialized form of reforestation. Research has shown that the land taken out of production to plant windbreaks in farm fields is more than made up for by increases in crop yield, if the windbreak is properly designed; planting windbreaks on farms is like putting money in your pocket. Similarly, windbreaks can result in substantial energy savings in your home by moderating winter winds.

One of the main principles of conserving biodiversity is to connect fragmented woodlots with corridors of natural vegetation. Vegetation buffers are corridors of vegetation along watercourses and around wetlands and ponds. They have the same benefits for wildlife as corridors elsewhere, but the benefit here is enhanced because aquatic habitats tend to be so biologically productive. At the same time, these buffers help improve water quality. If they are in the right location, they can also function as windbreaks.

Windbreaks bring many benefits; properly designed they can
- improve the yield and quality of crops,
- help prevent soil erosion and conserve soil moisture,
- save energy costs in buildings,
- provide cooling during hot summer weather,
- improve the productivity of livestock,
- provide wildlife habitat,

Windbreaks can control soil erosion and save up to 25% of the cost of home heating.

- influence snow distribution, and
- provide eventual timber or maple-syrup crops.

Windbreaks in agricultural fields have been shown to increase crop yields, in some cases up to 25 percent, which more than pays for the space they occupy in the field. Windbreaks around your home can save up to 25 percent of heating costs, while also providing cooling shelter in summer.

Similarly, buffers along watercourses or around wetlands provide multiple benefits. They can

- improve water quality and quantity,
- control soil erosion,
- provide cooling shade for fish habitat,
- filter out sediment and excess nutrients,
- provide wildlife habitat,
- provide insect habitat (in turn a source of food for fish), and
- if planted with appropriate trees, these buffers can provide an eventual timber crop or source of maple syrup.

Design of a Windbreak

The design of a windbreak depends on its purpose. Depending on the species used and the number of rows planted, windbreaks range from very dense to quite open. Higher densities are good as living snow fences and to protect cattle; medium densities are best for energy conservation around a home; and lower densities are ideal for distributing snow evenly across fields, reducing soil erosion and improving soil moisture.

In all cases, windbreaks should be planted on the windward side of the area to be protected. The choice of species also depends on soil types, as it does with any reforestation. Refer to Table 16 in Chapter 7 for choices of trees by soil condition. In most windbreaks, coniferous trees are used to provide the density required. Sometimes hardwoods or shrubs for wildlife can be added, if desired.

Eastern white cedar is the best species for a high-density windbreak. If used as a snow fence, it should be planted about 60 to 70 feet (20 m) from the area to be protected; most snow will accumulate in the first 50 feet (15 m).

Single rows of spruce or pine create general-purpose, medium-density field windbreaks, providing soil-erosion control and protecting crops. Medium-density windbreaks protect the largest areas; these are the most commonly planted windbreaks. On the other hand, if you want to spread out snow distribution in a field to improve soil moisture, a single row of hardwoods provides an appropriate low-density windbreak. In this case, you also need a row of shrubs to fill in the lower gaps as the deciduous trees grow taller. Silver maple, green ash, and poplar are among the hardwood species used.

In farm fields, a recommended rule of thumb is to have a tall windbreak every 220 yards (200 m) or a short windbreak every 110 yards (100 m), planted at right angles to the prevailing wind.

Single rows of spruce or pine about 100 feet (30 m) from a building will provide a shelterbelt for protection from cold winter winds. Be sure to make the windbreak long enough to extend past both ends of the building to be protected.

If your interest is in creating wildlife corridors, multiple rows of trees of different species, together with shrubs chosen to provide eventual food sources for wildlife, will be best. You can also mix in hardwood trees to provide habitat that is as natural as possible.

If your woodland has an abrupt vertical edge adjoining the farm fields, it is also beneficial to plant a narrow buffer of coniferous trees around the woods. This provides some protection from drying winds and improves the interior habitat, while adding variety.

Finally, you can easily enrich existing windbreaks or fencerows. Many farms have fencerows with at least a few large remnant trees along them. By simply stopping cultivation of adjacent land you can allow the fencerow to gradually widen to provide an improved wildlife corridor. Lightly plowing a strip to expose some mineral soil will enhance the natural succession. If there are sufficient trees in the existing fencerow, you do not need to plant new trees at all. On the other hand, you can plant additional trees if you wish – either conifers or hardwoods – as well as shrubs for wildlife food, along any existing fencerow. Planting a line of spruce or pine beside a largely deciduous fencerow will improve its function in controlling soil erosion. Consult Table 12 in Chapter 5 for a list of trees and shrubs you can use to provide food for wildlife.

Planting and Maintenance of a Windbreak

All the comments made in the previous chapter on reforestation also apply to planting trees for a windbreak. Trees must be chosen to match soil conditions, the ground must be prepared in advance, and the trees will need tending after planting. Survival rates will be heavily influenced by the care taken in handling and planting seedlings. Trees and rows are usually planted from 4 to 10 feet (1.2–3 m) apart, depending on the size of the trees and your plans for tending; you may want to mow between rows. Review the description of tree planting in Chapter 7 before you plan your windbreak planting.

Similarly, trees in any windbreak should be thinned as they grow, just as in coniferous plantations. You should aim to keep the crowns of adjacent trees just touching; remove every other tree when adjacent trees begin to touch. A few years later you will need to remove every other tree again until perhaps only one in four trees will remain at maturity. On average, you should aim to maintain a density of 40 to 60 percent in a medium-density windbreak; thin or replant regularly in order to achieve this.

Buffers along Watercourses

Buffers have the same wildlife benefits as windbreaks and can be planted the same way, but their real benefit is to water quality. They can minimize the runoff of sediments, fertilizer, and herbicides into streams, ponds, and wetlands. At the same time, they provide valuable habitat that maximizes the wildlife benefit of the aquatic ecosystem.

The zone of land along watercourses or around wetlands and ponds where dry upland habitat meets an aquatic ecosystem is called the "riparian zone," This is a specific element of the landscape that has high biological productivity because of the species

Allowing for a buffer of natural vegetation along all streams to protect water quality is one of the most important priorities for reforestation.

that grow there. The constant presence of water means that growth potential is rarely limited by lack of soil moisture, as is usual in upland habitats.

An inventory of wildlife will usually show that riparian habitats support a diversity of species not present elsewhere on your land. It is vital to remember that many such species (from ducks to dragonflies) use both the aquatic habitat and the surrounding riparian habitat as part of their life cycles. Therefore, a wide buffer of natural vegetation makes the habitats function efficiently. A wetland, pond, or stream with no buffer is severely limited in the species it can support. A buffer of 100 feet (30 m) or more is preferable, with wider buffers being better for wildlife. Ducks, for example, will often nest up to 1,000 feet (300 m) from a wetland.

Buffers also remove sediments, contaminants, and nutrients before they reach the water. This is a critical function in farming operations if agriculture and fish habitats are to be compatible. Even a narrow buffer of a few yards has been shown to have significant beneficial impact on water quality, but a buffer of at least 50 feet (15 m) is desirable for this purpose. In fact, buffers should vary in width beyond this minimum in order to extend out to the top of any slopes located beyond their average width.

In an active farm field, a buffer planted with permanent grass cover may be best, to enable the movement of farm machinery. But from our woodlot-management perspective, these watercourse buffers would be better if they were planted with trees. Both conifers and hardwoods, as well as shrubs that will provide food for wildlife, can be used.

In cases where erosion is occurring along stream banks, it may be possible to plant vegetation to control the erosion.Willows are ideal for stabilizing streambanks, due to their rapid growth and extensive root systems. Small rooted willow cuttings are available at many nurseries; they should be planted in the spring, along the bank just down to the high-water mark. If severe erosion is occurring and the bank is steep, you should consult a local water-control agency for advice.

Stream rehabilitation projects are popular with community environmental groups, especially fishing clubs. Many landowners have had stream rehabilitation work carried out on their own lands by cooperating with local community groups. It is well worth inquiring if you see a need for this on your land.

We hardly need to repeat that all our previous points about tree planting also apply in the case of stream or wetland buffers, as they applied to the planting of windbreaks. Review Chapter 7 to go over these details.

Planting Hardwood-Crop Trees

There are several variations on the use of deciduous plantings in reforestation that are more specialized (and demanding) than described in Chapter 7. We refer here to the specific choice of planting hardwood trees in an agricultural context as an eventual timber crop. This could involve planting selected trees in fencerows or buffers, in small areas of retired marginal land, or even in your backyard. It usually implies an intensive level of management, including pruning trees to improve the eventual quality of timber produced.

Planting hardwoods in the open and persuading them to grow tall and straight requires an understanding of a tree's response to light. Growing in a forest, trees will reach skyward, competing with their neighbors for sunlight; those that lose the competition will stagnate in growth or die. The degree of crowding required varies with the age of the tree. Very young saplings may grow in a dense patch with many others, at a rate of up to 50,000 per acre, so when these saplings are still an inch or less in diameter, they may already be 15 or 20 feet (4.5–6 m) tall.

Many of these saplings will die in the first few years, and all but a few will die before reaching maturity. During the long process of growth, trees compete with each other as they reach upward, eventually resulting in trees with tall trunks and few lower branches, which are ideal for timber. Spacing between individual trees expands over time with this growth, from a few inches in dense stands of saplings to many feet at maturity.

On the other hand, hardwood trees planted in the open do not have such competition to force them skyward, especially when planted at regular spacings of 6 to 8 feet (1.8–2.5 m) apart. Therefore, they develop low side branches and are unsuitable as sawlogs. The leader often seems to wander sideways, get nipped off by deer, or otherwise become crooked. Pruning and sometimes staking are required to ensure that such trees grow straight and tall. Tree shelters as described in Chapter 7 help hardwood seedlings get a good start on such growth.

If you are establishing hardwood species in a reforestation project in order to create a diverse natural ecosystem or wildlife habitat, their open-growth pattern does not matter much, but establishing hardwoods that you hope to harvest for timber takes a great deal more effort and maintenance. It is very costly, especially in terms of time, and is usually only considered on a small scale. As yet there is little successful experience to go by; however, it is our opinion that more landowners should consider this option on a small scale. Because of the effort that will be required to care for such trees, it makes sense to grow only those that will make valuable timber. Black walnut is perhaps the best example of this.

In many countries elsewhere in the world, trees in windbreaks or fencerows are routinely treated as crop trees. In New Zealand, windbreaks are commonly planted with two rows, one a slow-growing species and the other a fast-growing pine. The row of pines is pruned and eventually harvested for timber, and then replanted. Meanwhile, the other row continues to provide the windbreak function. Such arrangements are perfectly feasible in North America as well.

High-value hardwoods, such as walnut, black cherry, oak, maple, and ash, can be planted together with conifers along windbreaks, or to enrich existing fencerows. They can be planted on small patches of land to be retired from agriculture because of slopes or soil conditions, or they can be planted as a row of trees along a roadside or your driveway. You can select trees that are growing naturally as old fields come up in young hardwoods, or plant hardwoods into an field undergoing natural succession.

Even if you never plan to harvest these trees, having them can be like depositing money in the bank and accumulating interest. This could be done on small lots in the country, especially if you use the more valuable species such as walnut or black cherry.

Pruning Crop Trees

Pruning should start when these crop trees are quite young. The first task is to ensure that a straight leader grows upward. If the leading shoots of the seedling get nipped off or broken, the best remaining shoot needs to be chosen and encouraged to grow upward by careful pruning of competing leaders and staking of the chosen twig. In most cases, the unevenness of the trunk will eventually disappear as the tree grows larger. You can gradually prune away the lower branches to encourage upward growth while the tree is very young, but a good rule of thumb is never to prune away more than one-third of the foliage. As mentioned above, using tree shelters to help the seedlings is a good alternative to such pruning.

As trees pass the height of a tree shelter, or grow beyond 3 feet (90 cm) in height, continue pruning away the lower branches, preferably while they are quite small. Your goal is to eventually have a straight trunk that is at least 18 feet (5.5 m) tall with no branches. Try to prune branches while they are still under 1 inch (2.5 cm) in diameter, so you do not leave a large wound, but do not worry too much about small kinks in the trunk; in forty years these will have disappeared.

Judging by the natural growth pattern of young trees that come up in old fields that are undergoing natural succession, it would be desirable to plant hardwood seedlings in small clumps quite close together, so they will compete with each other for light and grow taller. Unfortunately, this means that only one will likely survive in the long run, so it would considerably increase the cost of planting. However, it is not necessary to thin out nearby trees until the crop trees are quite large; the nearby trees are actually an advantage in early years, encouraging the upward growth of the younger ones. Only when the crop trees get above 4 inches (10 cm) in diameter is it useful to thin out nearby trees. In this case, you should thin just enough to give the chosen crop trees some space on two sides of their crowns.

You can also prune conifers, such as pine, that are growing in plantations. You can start when the trees pass about 15 or 20 feet (4.5 or 6 m) in height, and the lower branches are beginning to die anyway. Take off the lowest branches, but never remove more than one-third of the living needles. By the time trees are about 25 or 30 feet (7 or 9 m) tall, and the plantation is ready to be thinned for the first time, you may have pruned some trees to about 10 feet (3 m). Your goal is to prune the first 18 feet (5.5 m) or so to create one knot-free sawlog in height. Such work is labor intensive and may never pay you for your time, but if you choose to do this, you will certainly improve the quality of the eventual crop of utility poles or sawlogs.

Right from the start, choose only the trees that will eventually be the best crop trees; you may want to follow forestry practices and mark the best with a blue dot to indicate they are your selected crop trees. Don't waste time pruning trees that will eventually be thinned out anyway, and remember that every other row of trees will probably be included in this.

All this is much easier said than done, especially with open-grown hardwood saplings. Trees seem to have their own stubborn independent growth pattern, branching just where you don't want them to. It requires an intensive commitment to pruning to ensure that you end up with straight, knot-free trunks.

However, if you put the effort into this, eventual financial returns should be promising. A single high-quality walnut or cherry tree could bring up to several thousand dollars; therefore, a small orchard or a fencerow of only fifty trees would be a good investment.

Maple orchards of sugar-maple trees planted for future use in maple-syrup production are another option for specialized hardwood planting. In this case, the open-growth form of the trees will not matter so much; indeed, the purpose is to create large open crowns on the trees, to maximize sap production and the sugar content of the sap. Maple trees should be spaced about 35 feet (10 m) apart and allowed to grow large wide crowns on short trunks. You should still prune trees during their early growth in order to create a branch-free work area around and between the tree trunks. Don't wait until the trees are too large to do this, or you will create large wounds. Trees can be tapped when they reach 10 inches (25 cm) in diameter, which may occur in as little as twenty to thirty years on high-quality soils.

Even two or three large sugar maples on a country property can provide enough syrup for home use over the year. Planting a few maples on any residential lot will eventually enable family maple-syrup production. This is a small aspect of woodland stewardship than can even be practiced in urban areas and makes a great hobby for retirement.

Christmas Trees

Christmas trees are a specialized forestry product that requires careful planning, lots of hard work, and product marketing, though it takes only a relatively small financial investment to get started on a small scale. Pine, spruce, and fir are the commonly used species. Scots pine is no longer recommended for reforestation because it develops such a crooked trunk when mature, but it is still a popular Christmas tree and can be well pruned to the right shape.

Christmas trees are an alternative choice for the use of old agricultural fields, rather than conventional reforestation, and they provide some wildlife habitat. On a smaller scale, it is certainly possible for any rural landowner to grow a few Christmas trees for their family and friends. A Christmas-tree plantation requires intensive management at particular times of the year, but it also brings a significant financial return. The first chapter in a popular book on growing Christmas trees is entitled "Money Can Grow on Trees."

A first step is choosing which species to grow. Perhaps the most common are Scots pine, white spruce, and balsam fir, but other species of pine, spruce, and fir are widely grown as well. Seedlings are available from Christmas-tree nurseries and sometimes from government nurseries. Check with any government nursery to see whether it allows seedlings provided for reforestation to later be sold as Christmas trees; some do not. You can also grow trees from seed, if you wish to take the time and trouble and have the necessary gardening skill. Most Christmas-tree growers initially plant small seedlings (less than 12 inches [30 cm] tall) in a transplant bed, like an intensive garden, for a year or two to give them time to develop strong root systems and grow a little taller before moving them to the open field.

Planting Christmas trees requires the same preparation and planning as would be the case in any reforestation project. The species used can grow acceptably on relatively poor soil, but not in wet soil. Planting is done as in reforestation, by spacing trees about 5 feet (1.5 m) apart in all directions, assuming you are planning to grow average-size Christmas trees. Using marked strings and planting rows with exact spacing will enable you to mow in more than one direction between trees for weed control. You should also leave two or three rows empty every fifteen or twenty rows to function as service roads. Site preparation is required, just as in reforestion. Reread the section on tree planting in Chapter 7. As with any reforestation, the most critical factor is to ensure that none of the seedlings' roots dry out.

Tending of Christmas trees is not as critical as with hardwood trees, but it is common to put more effort into it than with a simple reforestation plantation, since it is an economic crop. The most common practice is to prepare the site by using a chemical spray to kill the vegetation in strips where trees will be planted. Grass is allowed to grow between the rows of trees, but should be kept mowed. With a little careful tending, you can reduce the time it takes for the young tree to grow above the height of surrounding grasses, reducing your time between crops.

It may be useful, especially in the first year, to water your Christmas trees. Conifers have only a short active growing season of three to four weeks in the spring; if they miss this, they miss the entire year. With the cooperation of mother nature, you can plant early and the rain will get your trees started well. But if drought ensues and you are operating on a small scale, it may be well worth watering the trees during those first critical weeks.

Many Christmas-tree growers use fertilizers to enhance tree growth. However, the economic return on the use of fertilizers in Christmas-tree plantations varies with the species and the soil, and may not be worth the expense. Scots pine, in particular, does not respond well to fertilization, since it is adapted to grow on poor soil; in fact, higher fertility rates will slow its growth. White spruce and balsam fir will respond moderately to a low-level fertilizer applied every other year. White pine and blue spruce will usually respond well to fertilization. Beyond this, the fertilizer is likely wasted. If soils are already relatively fertile, fertilizer applications are really not warranted, though they may improve tree color when used in the last year of two before harvest.

Conifers usually require nitrogen more than other fertilizer elements, so if you do choose to fertilize, use a mix that emphasizes nitrogen (for example, a 10-5-5 mix), but not a super-high-level fertilizer, which can burn the tree. Fertilizer is most commonly applied in a circle just outside the outer branches (to encourage root growth), and is most easily applied on a small scale with a backpack spreader that has a release tube. Fertilizer can also be spread with conventional agricultural or home-garden equipment. You should consult your fertilizer dealer about the different types of fertilizers available and the equipment needed to apply them.

Be careful not to apply too much fertilizer; excess nutrients in the soil will draw water away from soil roots and possibly cause foliage browning or even death. Avoid spreading fertilizer near the trunks of seedlings. About 0.25 to 0.5 ounces (7–14 g) of elemental nitrogen per tree is recommended, and should be applied every other year, but not in the year of planting. To measure the amount of fertilizer you will require per tree, you may need to convert this into a volume measurement, which will vary with the type of fertilizer used. Consult your dealer for assistance, but with any form of fertilizer this will amount to less than 1 cubic inch (15 cubic cm) per tree – a very light application rate.

Fertilizers can cause serious water pollution if applied in high concentrations or if washed into streams by rainfall immediately following application. Avoid spreading fertilizer when rain is forecast. Some forms of fertilizers are quite dangerous to store; be sure

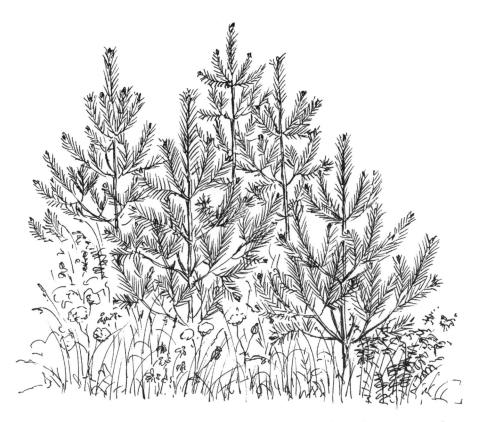

Any rural landowner can grow a few Christmas trees for his or her own use and relatives may relish the chance to cut their own tree.

to consult your dealer for safe storage procedures or try to buy only
the amount you will need when you need it.

Shearing and Pruning Your Christmas Tree

As your Christmas trees grow, the next task is to shear and prune
them. Shearing is the trimming of the year's new foliage to encourage the tree to become thicker, fuller, and the desired shape. To
accomplish this correctly, it must be done in the middle of the
trees's short spring growing season, after the season's new shoot
tips are 2 to 3 inches (5–7 cm) long. If new growth is sheared at the
correct time, while growth is still occurring, the tree will respond
by developing new buds between existing branches and filling in
with the desired thick foliage. Shearing at the wrong time will not
have this effect and may be hard to repair.

Shearing is usually done with a very sharp shearing knife,
hedge trimmers, or gas-powered or electric shearing tools. Some
have blades as long as the height of the tree. Shear from top to bottom in one sweep, working your way around the tree once; shape
the tree to the natural shape of the species, a thin upside-down
triangle for spruce and fir, and a slightly bulging shape for pine.

Pruning refers to cutting off woody growth. This usually occurs
with Christmas trees when extra leader shoots develop or after
trees are about 4 feet (1.2 m) high, at the same time basal pruning
is done. Basal pruning is the removal of thin lower branches to
open up a clear trunk for 4 to 6 inches (10–15 cm), or to the first
healthy whorl of branches. This encourages the growth of higher
branches and prepares the tree for harvest. Pruning shears, lop-
pers, or small saws can be used for this purpose.

Pests and Your Christmas Trees

Christmas-tree growers are apt to react quickly when a tree shows
signs of disease or insect pests. There are several things you can do

to deal with pests and disease; the first is not to panic. Most of the pests you will encounter are naturally present in small numbers in the environment, and will likely stay in those numbers.

Mechanical removal is probably the easiest means of dealing with a small infestation; even the loss of an entire tree is not a serious problem. Inspect your plantation regularly to catch problems while they are still restricted to individual trees.

You should also check carefully for any environmental stress. Trees often show symptoms such as discolored foliage because of moisture stress or lack of a particular element in the soil. Detailed soil tests are worthwhile to determine whether this is the case. If symptoms cannot be related to environmental factors and pests spread beyond one tree, consultation with a professional is in order. In this case, you should be able to see the actual insect pests, eggs, or larvae on your trees, or in the case of disease, the spores, rust, or mold.

Carrying a hand lens for examining trees and buying a copy of the *Christmas Tree Pest Manual,* published by the U.S. Department of Agriculture, may be worthwhile. Government extension services offer assistance with identifying problems and deciding on treatment. There are numerous chemical treatments available, but they must be used with caution and based on expert diagnosis.

Do not be fooled by the needles that turn golden and fall off the inside branches of some species. I remember one spring when a good friend was convinced that all his pine trees were dying. In fact, the white pine that he was growing lose their older inner needles on a regular basis; his trees were perfectly healthy.

The best protection against disease and insect pests is a diverse natural environment, which will attract predators and provide a more natural balance of nature. Unfortunately, Christmas-tree plantations, by their very nature, are not diverse. However, you can mix species together in rows or patches, plant fencerows of deciduous shrubs that will encourage wildlife, and leave standing dead trees as a perch for birds.

Marketing Your Christmas Trees

When marketing Christmas trees, you have a choice of wholesale or retail sales. Wholesale sales are usually completed early, by mid summer, and are based on written contracts, with a deposit from the buyer. There are many details to watch for, from having enough manpower to complete the harvest of trees as late as possible, to ensuring that truckers have the correct licenses if you are shipping trees across state borders.

Direct retail sales offer other challenges. You can set up your own Christmas-tree lot, or you can run your operation as a cut-your-own-tree outlet. This requires parking space, advertising, and a host of other arrangements. Either approach can bring you a healthy income, especially if this is a hobby.

It pays to plan your Christmas-tree planting as an ongoing operation, so you will have a steady supply of trees over the years. Planting a tenth of your available land in trees every year or every other year to establish a rotation of trees is a good idea. You can also interplant between trees as they are cut down, renewing your crop that way.

Both government extension-service professionals and Christmas-tree-grower associations can provide you with contacts, advice, and other sources of help. Make use of them as much as you need to.

Making Maple Syrup

Making maple syrup is a unique experience. From late February into late March, depending on your location, warm days and frosty nights will signal the arrival of maple-sugar season. Taps are drilled and buckets hung or tubing run to gather the sap. The sugar shack is ready with firewood, and a month of intense work begins for those who own a sugar bush.

You can taste the slight sweetness in the sap from the trees, but when it is boiled down for several hours to become maple syrup, the sweetness is indeed a miracle. Boil it a little longer and you get maple sugar, a treat so tempting it is hard to believe it came from a tree.

Maple-syrup production begins with woodlot management. Although some will suggest that you can get syrup from the sap of other maples, or even other species of trees, sugar maples are the only trees used in the industry. The ideal sugar-maple tree for sap production is a mature tree with a large healthy crown. Roadside or fencerow maples with their open-growth forms are ideal, and typically not only produce more sap, but sap with a higher sugar content.

In the woodlot, this suggests that thinning further than normal, to produce trees with larger crowns, would be desirable, rather than trees with tall straight trunks. Opinion is divided on how far this management for maple-syrup production should be taken. Some suggest gradually removing other species of trees and keeping only the best maple-crop trees; others suggest that pure sugar-maple stands will be more susceptible to disease and pests. Some natural diversity is desirable. No one disputes that a sugar bush must have some young maples to gradually replace large old tap trees.

Thinning is done to release the crowns of selected trees. Select the healthiest trees with the largest crowns as the crop trees to be protected. Thin some of the surrounding trees to open up the edges of these tree crowns by a few feet, but never thin a maple bush too intensively at one time. Growth will close these gaps in a few years. Larger openings will allow for regeneration of young trees.

Ideally, a sugar bush should have enough trees to allow for seventy to ninety tapholes per acre (175–225 per hectare); more than this means that you are protecting too many old large trees and may not have enough young regeneration growing. The number of trees will vary with the age of the maple stand, from only about twenty per acre (fifty per hectare) in very old stands of large trees to nearly eighty per acre (195 per hectare) in young stands just large enough to tap. The minimum size is conventionally 10 inches (25 cm) in diameter.

Tapping Sugar-Maple Trees

Tapping is done so you will be ready for the maple-syrup season, which varies from mid February in the southern end of the sugar-maple tree's range to late March in the north. The best runs of sap occur on days when the overnight temperature drops well below freezing, and then the daytime temperature rises quickly to several degrees above freezing. When this happens, you must be ready.

Sap can be gathered with buckets or with plastic tubing. In each case, a ⁷⁄₁₆-inch (1 cm) hole has to be drilled into the tree, about 3 inches (7 cm) deep. The actual height of the hole on the tree does not matter. The "spile" is securely tapped into the hole, and the tube is attached or the bucket hung. Large operations today almost always use tubing to gather the syrup; it saves an enormous amount of labor. Some operations are now using vacuum systems in conjunction with tubing, to increase sap yield. Large operations may involve up to several thousand trees, so tapping cannot wait until the last minute; in these cases, taps and tubing may be placed in the woods several weeks early. Large operators use specially adapted chain saws to power a drill for tapping.

It is critical to keep all the equipment clean, including the taps and buckets. Growth of microorganisms is the major cause of decline in the quality of syrup.

When the sap runs, it is time to boil it down into syrup. With miles of tubing running through your sugar bush, the sap will arrive right at the sugar shack for you; otherwise, the major work is in gathering the buckets. Modern evaporators use a system of small channels that carry the syrup over a roaring fire, in order to boil it down. The small channels maximize the heat contact with the syrup, and the wide-open pan maximizes the surface area for the evaporation of water. It usually takes about forty parts of sap to make one part of syrup, so a lot of water must boil off. One rule of thumb is to expect every taphole to yield about ten gallons of sap, which will provide one quart of syrup when boiled down.

The boiling process is straightforward until the sap finally reaches the point where it becomes syrup (when much of the water has evaporated). In a large evaporator this may only take forty-five minutes, but in a slow boil on a kitchen stove, it may take several hours. In any case, when the syrup is finally ready, the nature of the boiling changes subtly. The normal rolling boil of water is replaced with a surface of fine small bubbles. They will gradually cover the surface of the pan, and then suddenly – and we mean suddenly – they will boil up, and if you are not watching closely, the syrup will boil over the sides of the pan. The trick to making the best syrup is to catch the syrup just before this happens.

Professional sap makers use a hydrometer, but an amateur can use a candy thermometer to check the syrup. Sap, which is mostly water, boils at the same temperature as water, but as the water evaporates, the boiling point gradually rises until maple syrup boils at 7 degrees Fahrenheit (4°C) above the boiling point of water (212°F or 100°C at sea level). As soon as the syrup reaches this level, it is done. To check the boiling point of water at your location, stick your candy thermometer in a pot of boiling water beside your syrup (it also varies a little with the weather).

When your syrup has reached 7 degrees above that boiling point, take it off the stove. You should then filter the syrup by pouring it through a thick felt filter, and immediately pack it in clean jars for storage. Filtering will remove sediments known as sugar sand and result in a clear syrup product. You can buy special bottles or just use quart sealing jars. If the syrup has been properly boiled and is immediately bottled, it should store well.

A trick for when you see those tiny bubbles suddenly rising toward the top of your pot is to have a small amount of vegetable oil available. Just a tiny drop in the pot will take the bubbles back down safely.

Many amateur maple-syrup makers take their syrup into the kitchen for this last boil down. It is not a problem if the syrup cools down a little, then is reboiled on the kitchen stove.

Professional maple-syrup operators have to meet a number of guidelines and regulations governing the quality of their syrup before it can be sold. A grading system is in place in the U.S. and Canada, with quite strict regulations to prevent the watering-down of anything sold under the label "maple syrup." (Many commercial syrup products have only 1 or 2 percent real maple syrup, or none at all.)

There are active maple-syrup-producer associations in most of the northern states of the eastern deciduous-forest region, and in all the adjacent provinces of Canada. There are also a number of supply houses that provide the equipment needed for maple-syrup production. Although Vermont has a widespread reputation as the leading U.S. producer, Quebec produces more than all the other top-ten producing states or provinces put together, including Vermont.

Commercial-scale operations require major investments of time and equipment, but anyone with even two or three large sugar-maple trees can experience the magic of making maple syrup. For several years, our family's supply of maple syrup came from three large old maple trees in our front yard near downtown Guelph. Find a row of old maples along the roadside, and you will have more sap than you can cope with for home boiling. Several months later, it will make those morning pancakes taste great.

Further Reading

Agriculture Canada. *Best Management Practices: Farm Forestry and Habitat Management.* Toronto: Ontario Federation of Agriculture, 1994.

———. *Best Management Practices: Fish and Wildlife Habitat Management.* Toronto: Ontario Federation of Agriculture, 1996.

Benyus, Janine M. *The Christmas Tree Pest Manual.* Washington: U.S. Department of Agriculture, 1983.

Coons, C.F. *Sugar Bush Management for Maple Syrup Producers.* Toronto: Ministry of Natural Resources, 1983.

Hill, Lewis. *Christmas Trees: Growing and Selling Trees, Wreaths and Greens.* Pownall, VT: Storey Communications, 1989.

Landowner Resource Centre. "The Benefits of Windbreaks" (fact sheet). Manotick, ON: Landowner Resource Centre, 1994.

———. "Designing and Caring for Windbreaks" (fact sheet). Manotick, ON: Landowner Resource Centre, 1994.

Lawrence, James, and Rux Martin. *Sweet Maple.* Shelburne, VT: Chapters Publishing, 1993.

Mann, Rick. *Backyard Sugarin'.* Woodstock, VT: Countryman Press, 1991.

McEvoy, Thomas J. *Using Fertilizers in the Culture of Christmas Trees.* Hinesburg, VT: Paragon Books, 1992.

CHAPTER 9

Trails, Pests, and Poachers

A ccess to your woods through a trail system is an important issue that complements all other management options. Here we outline why and how to design a trail or road system for your woods. We also deal briefly with two other issues: the presence of insect and animal pests and disease, and the problem of trespassers.

Woodland Trails and Roads

Whether your interest in woodland management is in observing nature or in harvesting timber, having an appropriate trail to follow is a big advantage. This is equally true for areas you have reforested, which you may eventually need access to for thinning. Landowners who have established walking trails on their land find that they get out and walk much more often, simply because their land is easily accessible. As a result, their knowledge and appreciation of their land increases. A properly designed trail will also minimize damage during timber harvesting.

Why do we emphasize trail building in this chapter? There are two reasons.

First, we have found through personal experience that landowners learn to appreciate their woodlands much more deeply if they regularly walk through them. It is our belief that, in the long

Clearing a proper trail in the woods will open a world of enjoyment to any woodland landowner making it easy to appreciate your woodland in all seasons.

run, this appreciation will lead to a greater commitment to conservation. Without a trail that makes walking easy, this doesn't happen. While most landowners will explore their woodlands occasionally, having to dodge branches and pick their way between saplings prevents many from doing this frequently, especially with family members who may be a little reluctant in the first place.

On the other hand, landowners who have trails that make walking easy report to us that they open up a whole world of enjoyment to them. They are comfortable taking guests and children on woodland walks, and they learn a great deal more about their land. Trails also make for great cross-country skiing in the winter. In the long run, we believe this will reinforce personal commitments to woodland conservation.

The second reason to build a trail is to harvest timber or firewood. In this case, having a properly designed tractor trail to follow will dramatically lessen the environmental impact of harvesting operations.

Although logging can be done carelessly and too intensively (as is discussed elsewhere in this book), the biggest environmental damage from timber harvesting usually comes not from cutting the trees, but from hauling them out of the woodland. There are many practices that can minimize this damage, such as logging only when the ground is frozen or wet, using horses, and winching logs out to trails. But perhaps the most important factor in minimizing negative impact is having a properly designed trail system. For timber hauling, a trail will have to be wider than a walking trail, of course, but taking care in designing the trail, will minimize damage when harvests are carried out.

If you are hesitant to build a trail or road through your woods, remember that it is not the trail that is disruptive, but the use made of it. As long as you design and use it appropriately, a trail will enhance your appreciation of your woods, with virtually no serious impact. It will also have very little effect, if any, on the wildlife in your woods.

We are talking in this case about narrow tractor-trail-width woodland roads, which are typical of private woodlots, not major construction for logging trucks, as may be required in large northern forestry operations.

Any stream that provides fish habitat may be protected by law; you will require proper legal permits to construct stream crossings and, in some cases, a preceding environmental-impact assessment. In some cases, you may require a permit for road building in your woods.

Principles for Trail and Road Layout

There are several principles for designing a trail in your woodland. All of these apply to simple walking trails and larger woodland roads alike; it is only the size of the trail (and the cost of construction) that differs.

1. Keep trails away from wet areas as much as possible, including patches of land that may only be wet during the spring. Most of the unsightly damage that occurs from trampling along hiking trails or from timber hauling occurs where trails cross wet patches of soil.

2. Where necessary, install protective structures, such as culverts, drainage pipes, or small bridges, to cross wet areas. These structures can be very simple, but will significantly reduce or eliminate trampling and compaction.

3. Avoid steep slopes as much as possible, especially with woodland roads, even if this leaves some areas of your woodland inaccessible. For walking trails, route the trail on a gentle angle across the slope, creating switchbacks if necessary.

4. Build your trail to avoid soil erosion, using water bars, log headers, rock berms, or long steps to help control runoff where necessary. Grade roads so they are higher

in the center and water runs off to the sides, not down the middle. Keep slopes very gentle.

5. On larger woodland roads, avoid routing ditches directly into streams; direct the flow of water to low spots where infiltration can occur. Protect streams and wetlands from sedimentation during construction and during logging operations with straw bales or filter barriers.

6. Design walking trails to be narrow, unobtrusive, and winding; situate them where they will pass by points of interest, but avoid sensitive features, such as nesting areas for rare species. If you intend to cross-country ski on the trails, avoid sharp bends and angles.

7. Design woodland roads to be only as wide as needed; keep bends to a minimum, and create gradual curves for easy timber hauling. Avoid clearing such a wide trail that you significantly open the canopy above the road.

8. To avoid trespass problems, situate the entrance to your trails where it is not visible from a road. Design your trail in a loop for walking, and install a bench halfway to rest on, perhaps at an interesting view-point. Place a locked gate across the entrance to a woodland road.

You should only plan your trail route after several woodland walks, preferably during all seasons. To establish a walking trail, become familiar with your woodland, then note areas that you want to avoid and points of interest you want to pass close to. Then follow the easiest walking route you can, preferably in a loop. If you can't avoid some wet areas or steep slopes, you may have to install some simple structures at these points, but in the long run, you will be much better off creating a trail that is a little longer if this will enable you to avoid the wet or steep areas. You may want to purchase some bright-colored flagging tape at a hardware store to lay out your trail, adjusting it until you are satisfied before doing any trimming.

For woodland roads, you must plan more carefully, watching particularly for steep slopes and wet areas. Long straight stretches or gradual curves with very gentle corners are best. Flag the route to be cleared.

Walking Trails

Woodland walking trails can usually be built by trimming small trees here and there, and trimming branches up to above your eye level. For most areas, walking the proposed trail route a few times will soon create the path that you need. Over the year, you may need to kick fallen branches out of the way or trim branches that intrude on your trail at eye level. In a few areas, you may need to take a shovel or dump some gravel to even out the terrain for walking.

Crossing wet areas or small streams can be done with simple bridges built with two logs and a deck of two-by-fours. Alternately, if the wet area is only a small line of seepage, you can bury common plastic field-drain pipe to carry away the water. Bury the plastic between two logs or two-by-fours to provide extra support and to keep the pipe straight; cover it with at least the same depth of soil or gravel.

On slopes, a variety of techniques can be used to make walking easier and minimize erosion. Run the trail across the slope as much as possible, and use log headers or rock berms to edge the trail. Build steps out of railway ties or logs with supporting stakes. Be careful to space the steps for comfortable walking. "Water bars" are shallow ditches built on an angle across the trail or logs buried across the trail so that they protrude only 1 or 2 inches (2–5 cm) above the ground; this will intercept water running down the trail and direct it to the outer edge, reducing trail erosion.

Walking trails can be created through old fields by mowing them. It is necessary to do this before succession proceeds too far, because once woody vegetation begins to grow, you can no longer

Control erosion on trails with water bars, log or rock headers, or steps.

mow a path. If you begin before this stage, you can maintain a path easily by mowing it three or four times a year. Wait as long as possible before mowing in the spring to avoid disturbing nesting birds.

Mowed trails through old fields are ideal for cross-country skiing, especially if you mow them once late in the fall. On the short grass of the trails only a little snow cover is needed to provide skiable paths.

Maintaining a mowed trail makes for easy walking through old fields.

Once you have built a woodland trail, use it. You'll find that your appreciation and understanding of your woodland increases the more regularly you walk through it.

Woodland Roads

Building woodland roads is a more expensive proposition, but all the same principles apply as building trails. Avoiding wet areas and steep slopes are the most important among these.

First, you should look for any existing old tractor trails. These may be hard to discern in the summer, especially if they are overgrown with young saplings; they are sometimes easier to see in the winter after a light snowfall. They make a very good initial layout for your own road system, and renewing them may be much cheaper than building new roads. If they were laid out originally for horses, they likely avoid steep slopes and wet areas that you should be circumventing anyway.

Since woodland roads may occasionally be used by heavy equipment, it is even more important to avoid wet areas and to design them carefully to protect drainage patterns. Large wet areas should be totally avoided; small wet patches may require a buried culvert to avoid damaging the terrain.

If you need to cross a flowing stream, you may need a permit to install a culvert or bridge. The law does not allow you to disturb a stream bed by removing gravel, adding gravel, or rerouting the stream. A buffer of undisturbed natural vegetation should be left along all streams, as well as around wetlands and ponds. Roads should cross streams where they are straight, not at bends, and should cross at right angles to the water flow.

Culverts should be installed close to right angles as well. They should be placed carefully at the right depth to capture water flow and should slope downward at least ¾ to 1 inch (2–3 cm) for every yard of culvert (a 2 to 3 percent slope). Water should drop slightly upon entering the culvert, so the culvert should be placed just below the stream-bed level; water should not drop upon leaving the culvert or it will cause erosion. The stream bed at both ends of the

water drops slightly entering culvert

at least one foot of cover, or thickness of culvert if less than one foot

rock armour for inlet and outlet

Proper installation of a culvert will minimize erosion at stream crossings.

culvert should be protected with rocks, and the ends of the culvert should extend out far enough from the road so that the backfill of the roadbed does not enter it.

In the case of any stream that flows regularly, you need to consult the experts, both for legal permits and for advice. It is also important to ensure that any culvert or bridge installed is large enough to handle peak flood flows (usually in the spring). Otherwise, it can easily be eroded out, causing greater damage to the stream. Be sure you visit the site in the spring, when water levels are their highest, before designing any stream crossing, and be sure you get the required permits.

In wet patches without a flowing stream, a drainage pipe or some logs with a surface layer of gravel may provide enough protection. Drainage pipes should be laid between two logs for strength and kept straight. You can also make wooden culverts from two logs and a wooden deck. The ultimate aim is to avoid degradation of water quality by preventing any sediment from entering flowing water, and to avoid deep ruts in your trail created by improper drainage. The most desirable practice is to avoid any use of woodland roads by machinery during wet weather; wait until the ground is dry or frozen, and you will eliminate 90 percent of the damage.

Slopes are the second cause of environmental damage. Water flows quickly down even a slight slope, causing erosion and carrying sediment with it. The worst erosion occurs during major storm events, perhaps only once every several years, when a road itself becomes a stream channel. The power of running water can do enormous damage. Without proper design, this can require costly repair and damage nearby streams.

Road and trail slopes are best kept under a 6 percent slope, and preferably only 2 percent; that is, they should not rise or fall more than 22 inches (60 cm) in 30 feet (10 m) of distance, or preferably 7 inches (20 cm) in 30 feet (10 m). Anything above this slope will require special and more costly construction techniques, including engineered slopes and ditches to prevent erosion.

Designing your woodland road to avoid such slopes in the first place will save you significant repair costs in the long run. Winching logs off slopes can compensate for this during timber harvests.

In a small woodlot, you may need nothing more than a tractor trail, which you can create by clearing saplings and a few trees, while protecting your wet areas. In larger areas, some construction with a bulldozer may be necessary. In this case, roads should be graded in a convex shape, making the center higher so the road itself does not function as a ditch during rainfall events. Standing water on a road and rutting in the road are always signs of poor drainage. Never let ditches along the side of a road empty directly into a stream.

Adding gravel dramatically increases the cost of a road, but a little gravel in selected locations can minimize erosion problems. Gravel is much more resistant to erosion than bare woodland soil.

When constructing a woodland road, you can do the layout and design yourself, and proceed with most of the road clearing. Then you can hire a local contractor, who can bring in a small bulldozer to do the rest. A skilled operator who has experience in such work is essential. If you will require some bulldozer work, remember to leave stumps of larger cleared trees about 3 feet (90 cm) high, to give the bulldozer operator the leverage to pull them out.

Roads require maintenance. Walk your roads, clear off fallen branches, and keep overhanging limbs pruned back. Pretend you are riding a horse when pruning; prune high enough so you don't unexpectedly get a branch in the face later. Clear out the entrances and exits of all culverts, and repair any erosion at the ends. If necessary, mowing the roads annually with a bush hog will keep new woody growth down. After logging, have the roads back-bladed smooth again, and have any ruts repaired.

Insect Pests and Diseases of the Woodlands

Forest owners are often worried about pests and disease. There are numerous insects that can attact trees, causing defoliation, cankers, or eventual death. A few of these can cause serious, widespread damage, while others create very minor, widely scattered infestations.

By far the most damaging tree pest in eastern North America is the spruce budworm; however, it occurs primarily in regions north of the hardwood forest, among the conifers (especially spruce) of the mixed and boreal forests. Outbreaks have been severe enough to cause widespread tree mortality in some regions, and aerial spraying has been undertaken to control the budworm in target areas, such as high value forests and parks. It is not a major pest in the hardwood forests further south.

Other major pests, such as the jack pine budworm, the jack pine sawfly, and the Bruce spanworm, also tend to attack coniferous trees in the northern regions of eastern North America. In the deciduous hardwood forests that this book addresses, the forest tent caterpillar is perhaps the most serious pest. Other important pests, though they occur primarily in local outbreaks, include the hemlock looper and the oak leafshredder. The gypsy moth can cause widespread tree defoliation during its outbreaks, but there is no evidence that it causes tree mortality.

Dealing with such pests first requires correct identification of the problem. Samples of insects, or of damaged tree branches, leaves, or needles, can usually be identified by a professional or by an extension forester.

Do not overreact to evidence of disease or insect damage on your trees. Most pests occur naturally in cycles and are controlled through natural ecological checks and balances, as are other insects and birds. Especially in a small area, you can manually remove any infestation you are worried about.

More serious infestations, and treating them with pesticides, will require the advice of a professional. There may also be strict rules about pesticide applications that you must follow.

Animals can be serious pests as well as insects and disease. Deer, rabbits, mice, voles, and porcupines can all cause extensive damage to reforested areas. The use of tree guards is probably the best protection against such pests. You might also want to consider encouraging hunting if wildlife damage is excessive. A wildlife biologist can provide further advice.

Absentee Land Ownership and Trespassers

Many private woodlots are owned by individuals who live elsewhere, and are left unsupervised for long periods of time. There are particular challenges in absentee land ownership that can cause you problems if you are not careful. These include general problems of trespass, hunting, bush parties, and poaching of timber or firewood.

There are several steps you can take to try to avoid these problems.

- Visit your property as often as possible; a regular human presence means that the property does not look unused.
- Keep the entrance tidy, making it look like someone maintains the site regularly.
- If you have a building on the property, keep it tidy and well cared for.
- Most important, get to know some local residents; let them know who you are and how to get in touch with you.

This last point is fundamental. If you are known by your neighbors and in the local community, people will watch out for you and your property. Rural communities are oriented to this. Also consider hiring a local resident to keep an eye on your land, perhaps allowing him to cut firewood or hiring him for construction or timber harvesting in return. Someone who feels the responsibility to keep an eye on things can be a valuable local ally.

It is also useful to develop arrangements with local residents for groups to make active use of your land, especially if you cannot often visit. Field-naturalists' clubs or snowmobile clubs might appreciate the chance to use your woodlot. Again, this gives these users some responsibility, and there will be more people watching over your land. With known users you can negotiate clear agreements over what is allowed and what is not.

One way to do this is to have an annual event, such as a walk in the woods or a "teddy-bear picnic". Even one annual event where local citizens feel welcome to visit your land can create a welcoming local attitude.

Hunters are often of particular concern. Landowners tend to have strong feelings either for or against hunting. In our experience, the best way to cope with trespassing hunters is to make a specific arrangement with a local hunter or group of hunters for the right to hunt on your land, in return for watching out for other trespassers. You may be rewarded with a little venison in return.

Opinion is divided on whether you should lock up your property to try to prevent access at all. A locked gate is quite appropriate in a built-up landscape within commuting range of urban areas, but in more distant rural areas this contradicts the advice we have given above. If you are encouraging a local group to use your land, they need access. On the other hand, if you are hiring a local resident as a guard, you can lock the gate and give him the key.

If you are a non-resident of a rural area, we do not advise you post your land with no-trespassing signs unless you have a specific problem with unacceptable uses. Such action is more likely to build an unfriendly barrier between you and your neighbors, just the opposite of what you want.

Further Reading

Armstrong, J.A., and W.G.H. Ives. *Forest Insect Pests in Canada*. Ottawa: Natural Resources Canada, 1995.

Ashbaugh, B.L., and R.F. Holmes. *Trail Planning and Layout*. New York: National Audubon Society, 1967.

Boyce, J.S. *Forest Pathology*. New York: McGraw Hill, 1961.

Demrow, C., and D. Salisbury. *The Complete Guide to Trail Building and Maintenance*. Boston: Appalachian Mountain Club Books, 1998.

Fazio, James R. *The Woodland Steward.* Moscow, ID: The Woodland Press, 1985.

Forest Service. *Trails Management Handbook.* Washington: U.S. Department of Agriculture, 1985.

Hepting, G.H. *Diseases of Forest and Shade Trees of the United States* (handbook #368). Washington: U.S. Department of Agriculture, 1971.

McGauley, B.H., and C.S. Kirby. *Common Pests of Trees in Ontario.* Toronto: Ministry of Natural Resources, 1991.

Myren, D.T. *Three Diseases of Eastern Canada.* Ottawa: Natural Resources Canada, 1994.

Tattar, T.A. *Diseases of Shade Trees.* New York: Academic Press, 1978.

Wiskel, Bruno. *Woodlot Management.* Edmonton, AB: Lone Pine, 1995.

CHAPTER 10

DEVELOPING YOUR WOODLAND- STEWARDSHIP PLAN

In spite of the fact that this book has dealt extensively with woodlands themselves – woodland ecology, steps for preparing a woodland inventory, and management options for woodlands – the most important feature of a stewardship plan is not the woodland, but you the landowner. Central to any stewardship plan must be your own personal interests, abilities, financial and family position, and goals for your land. It is based on these factors that your woodland-stewardship plan is constructed.

You may wonder why we have left the discussion of developing your own goals and objectives and preparing your stewardship plan to the end of this book. Many would say that this step should be undertaken first. We do so because in our experience most rural landowners, while very interested in learning more, do not know much about their own woodlands or about woodland-management options. We believe you should keep your mind open about how you wish to manage your woodland until you have learned about these options; then you are ready to consider what to do on your own land.

We present this discussion of how to prepare a stewardship plan as a series of simplified steps. There is no need to follow these exactly; they are merely a guide. If you are already very knowledgeable about your woodland and are actively managing it, you can skip through to preparing a plan. But if you are just getting started in caring for your woodland, these steps may help.

1. *Prepare a Draft Woodland Inventory*

The first step in preparing your plan is to do a preliminary, or draft, woodland inventory. You cannot begin thinking about the possibilities for your own woodland until you are at least generally familiar with it. If you have to learn to identify the trees, this can be a challenge at first; allow yourself at least a full growing season to walk through your woodland observing, identifying the trees, and perhaps counting some of them to establish the species composition of your woods.

As described in Chapter 5, you should walk your woods in all seasons, from early spring during the snow melt through to late autumn after the leaves have fallen. Watch particularly for drainage patterns, soil conditions, and for unique wildlife habitats. Purchase a field guide and learn to identify the trees through their leaves, and then practice identification by looking only at the bark. Of course, you might choose to hire a consultant to prepare an inventory for you, or find an expert biologist who can walk through the woods with you. We hope that Chapters 2, 3, and 4 will help you know what to look for.

This preliminary description should take you as far as preparing a rough sketch map of the ecological communities in your woods (see page 70) and noting a few points to help describe each community. You should also be able to list any special features.

2. Become Familiar with Management Options

Most of the rural landowners we have visited are largely unfamiliar with possible management options for their woodlands. If they are familiar with some options, it is usually timber management or firewood cutting, and even here, they do not understand the steps they should take to get started.

Reread Chapters 5 to 9 of this book. Don't worry which specific options apply to your own woodland; just become familiar with the possibilities. Don't rush into decisions; take the time to consider which management options are of interest to you.

We have tried to present the full range of management options, with a strong emphasis on environmental sustainability as a minimum choice. We have also emphasized understanding and conserving biodiversity, as well as more economically oriented options, such as timber management. This reflects the evolution of woodland stewardship and the science of forestry today, which takes a much more holistic view of the entire woodland ecosystem in deciding on management plans.

3. Carefully Consider Your Own Goals and Objectives

The single most important step is to establish your own personal goals and objectives. Think about your personal interests, the skills you have, the amount of work you wish to do, your financial position, and your family interests in order to establish your goals for your woodland. You are the key to these decisions. While the ecological characteristics of your woodland are important, and understanding management options is necessary, in the end it is the landowner who decides on a stewardship plan.

So take your time and consider these questions carefully.

You may be fascinated by birds and plants and want to keep a detailed journal of observations in your woodland, or you may be allergic to mosquitoes and too busy to spend much time there. You may own your woods primarily for personal pleasure and recreation, or you may be using it as a primary source of income. So you need to ask yourself a few key questions:

- What are my main interests in owning my woodland?
- How much time do I want to spend there?
- Do I want to just walk and enjoy my woodland, or do I want to work at cutting the firewood or planting trees?
- Do I need a regular financial return from my woodland?
- Am I interested in learning about the woods myself, or do I want to hire experts to help me?
- Am I interested in doing the management myself, or do I want to hire someone else to do it?
- Am I interested in joining other landowners with similar interests?

4. Choose Your Objectives

At this point you should be ready to choose your objectives for your woodland-stewardship plan. These should be based on your interests, skills, and abilities. They might include

- becoming more knowledgeable about your woods,
- protecting wildlife habitats and biodiversity,
- using the woods for recreation,
- cutting timber or firewood for sale,
- producing maple syrup,
- reforesting marginal land to control erosion,
- protecting water resources,
- documenting the ecology of your woodland, or
- seeking professional help to document any rare species.

It helps to make a list of these objectives and to then decide which objectives are most important. Most landowners have one or two main interests in owning their woodlands, and many smaller interests. You can combine these in any way you wish, although some are more compatible with each other than others. Usually, landowners find that even though they start with one main goal, as they become more knowledgeable, their interests expand.

In extreme cases, some activities may be incompatible. For example, intensive timber harvesting does not serve to conserve biodiversity. In our opinion, however, as long as you follow the minimum guidelines for environmental sustainability outlined in this book, most activities should be compatible. If there are difficult choices to be made, it is the landowner who should make those choices, based on understanding the ecology of his woods.

Your choices may be limited, however, if your woodland is a habitat for rare species, if it includes unique biodiversity features, or if it has significant old growth; then you should leave it primarily for conservation and avoid activities that disrupt these features.

You can also change your goals over time as you experiment with management options in your woods. In fact, scientists now use the term "adaptive ecosystem management" to indicate that management needs to adapt over time. You may want to try something on a small scale before trying a bigger project. You may also simply need time to prepare a more detailed inventory and become more knowedgeable in order to make further management decisions.

5. *Choose Your Management Options*

At this point, you can make specific choices for the management activities you would like to carry out. Keep in mind the time and ability that you are able to devote to this; consider hiring someone for certain activities, if necessary.

The choices of management activities you make should reflect the objectives you have chosen. For example, if you choose to conserve biodiversity and wildlife habitats, then your activities might include options such as

- conducting a detailed inventory of plants and wildlife,
- keeping a record of spring-wildflower, frog, and bird species,
- putting up birdhouses around the edge of your woodland,
- fencing out cattle,
- expanding your woodland by allowing natural succession,
- planting a connecting corridor of coniferous trees, or
- creating a walking trail.

If you have decided to harvest firewood and timber, your management activities could include

- a detailed timber inventory,
- professional advice to set a sustainable level of harvest,
- having your woodland professionally marked for cutting,
- the protection of snags, den trees, and cavity trees for wildlife,
- keeping a few supercanopy trees,
- designing skid trails to avoid wet areas and steep slopes,
- careful operations and cleanup by loggers, and
- building brush piles with leftover branches.

The list of options you may undertake is only limited by your imagination and by your ability. But don't plan to do too much; plans that are not implemented are only discouraging. After all, woodlands grow slowly, and management options should be thought of over decades, not days or weeks.

It is important to set a schedule for carrying out activities. Most government agencies set plans for five-year time periods, with a list of what activities will be done each year. But some aspects of your plan may only occur over twenty years. Remember, there is no hurry.

This is also probably the best time to consult a professional expert for assistance, if you haven't done so thus far. Even the dedicated landowner who prepares his own inventory and plan, and expects to do the work himself, will benefit by reviewing his decisions with an experienced professional. Your state or provincial forestry or natural-resource agencies, or non-government associations, should be able to give you a list of names to contact. Professional review is especially important if you think you may have unique wildlife or rare species in your woodland, or if you plan a timber harvest. A mistake at this stage can be very costly later. Professional consultants should be prepared to provide a brief consultation at no charge and should explain their fees for more detailed work.

6. Revisit and Finalize Your Woodland Inventory

Depending on your choice of management options, you may need to return to your woodland to do more inventory at this stage. If you have developed an interest in wildlife-habitat features, for example, you can go back and prepare a more detailed list of these. Similarly, if you have decided on timber harvesting, you need to prepare a detailed timber inventory, if you have not done so, or hire a consultant to do so. You may not be able to finalize your choice of management until this inventory is complete.

You should then end up with the detailed final map of your land, along with detailed descriptions of each ecological community and any special features. See Chapters 3 and 4 for an outline of what to include.

The level of detail covered in your final inventory should depend on the objectives you have chosen. If your chosen management activities reflect an interest in nature conservation, then there is no need to do a detailed timber inventory, but if you intend to harvest timber, you do need that information. Adapt your inventory to your own chosen purposes in managing your woodland.

7. *Write Down Your Stewardship Plan*

You may be asking, Why should I bother to write this all down? Many landowners keep a mental inventory of their woodlands, and have a mental plan for what they are going to do. This is the way we usually live; we do not write down plans for many of our activities. You may be very knowledgeable, visit your woodland in all seasons, and have a thorough understanding of management choices, but you keep this all in your head.

There are several advantages to writing down your plan

- Going through the steps of preparing a written plan will likely help you recognize options you were previously not aware of, resulting in more economic return and more effective environmental protection.

- As discussed later in Chapter 12, a written plan provides the basis for considering the long-term future of your woodland, if you plan to leave it to your heirs or sell it. There can be significant financial and conservation advantages through several legal tools, but they all depend on a written management plan.

- There can be significant tax advantages in treating your woodland as a business. Again, written documentation will likely be required.

- Most government support programs for woodland management require a written stewardship plan of some sort, and such programs may provide important financial incentives to do so.

- Discussions with professionals who may assist you in management or timber harvesting, or with logging contractors, are much easier if you have a written plan to refer to.

- Writing it down can help you learn, resulting in a deeper appreciation of your woodland and more enjoyment from your land ownership.

To help you understand what might be involved in deciding your options, we have prepared an example, presented in the following two pages. The summary we present here is a very simple one; depending on your interests, you may go far beyond this level of detail. This plan is based on the same property used as an example in Chapters 3 and 4, so you can compare the inventory maps and the aerial photo to this final plan.

SAMPLE MANAGEMENT PLAN

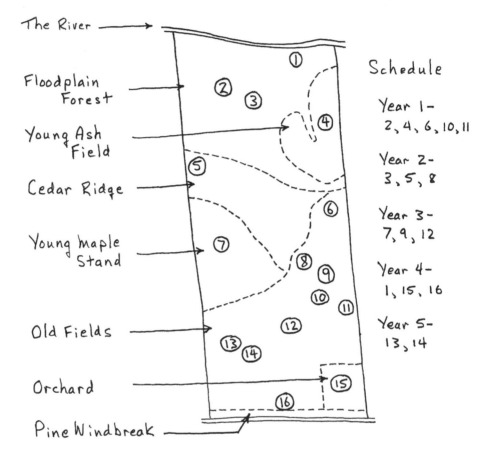

The River ⟶

Floodplain Forest ⟶

Young Ash Field ⟶

Cedar Ridge ⟶

Young Maple Stand ⟶

Old Fields ⟶

Orchard ⟶

Pine Windbreak ⟶

Schedule

Year 1–
2, 4, 6, 10, 11

Year 2–
3, 5, 8

Year 3–
7, 9, 12

Year 4–
1, 15, 16

Year 5–
13, 14

LANDOWNER OBJECTIVES

- Use for family recreation and education
- Develop pond for wildlife habitat
- Family scale maple syrup production
- Long term timber harvest potential

MANAGEMENT ACTIVITIES

1- Develop fishing access point.

2- Allow hunting by neighbours in return for venison.

3- Do timber inventory to evaluate for thinning.

4- Prune and thin ash for future timber.

5- Develop walking trail in woodland.

6- Maintain meadow and trails by mowing.

7- Select maples for future syrup production

8- Plan pond for wildlife habitat.

9- Build fort for kids.

10- Erect birdhouses with children.

11- Plant conifers along fencerow for privacy.

12- Plant diversity of tree species.

13- Build new larger shack for future family visits.

14- maintain views for future retirement home.

15- Improve orchard management.

16- Control invasive Scots pine.

Keep annual record of management.

Further Reading

Beattie, M., et al. *Working with Your Woodland: A Landowner's Guide.* Hanover, NH: University of New England Press, 1993.

Brett, R.M. *The Country Journal Woodlot Primer.* Brattleboro, VT: Country Journal Publ., 1983.

Decker, D.J., et al. *Wildlife and Timber from Private Lands: A Landowner's Guide to Planning.* Ithaca, NY: Cornell University Extension Service, 1983.

Fazio, J.R. *The Woodland Steward.* Moscow, ID: The Woodland Press, 1985.

Hilts, S.G., and P. Mitchell. *Caring for Your Land.* Guelph, ON: University of Guelph, 1994.

Hilts, S.G., and P. Mitchell. *Taking Stock: Preparing an Inventory of Your Woodland.* Guelph, ON: University of Guelph, 1997.

Loomis, R., and M. Wilkinson. *Wildwood: A Forest for the Future.* Gabriola, BC: Reflections, 1990.

McEvoy, T.J. *Introduction to Forest Ecology and Silviculture.* Burlington, VT: University of Vermont, 1995.

Minckler, L.S. *Woodland Ecology: Environmental Forestry for the Small Owner.* Syracuse, NY: University Press, 1980.

Walker, L.C. *Farming the Small Forest : A Guide for the Landowner.* San Fransisco: Miller Freeman, 1988.

Wiskel, Bruno. *Woodlot Management.* Edmonton, AB: Lone Pine, 1995.

CHAPTER 11

Buying Woodland Property

The mystique of owning your own piece of property in the country maintains a strong hold on many would-be landowners now locked in the confines of urban living. Perhaps it is due to a childhood spent growing up in the country, or perhaps visits to grandparents in a smaller town or on the farm. Whatever the reasons, owning land, and for many, living in the country, is an ongoing dream.

In this chapter, we provide some brief suggestions on how (and whether) you should go about buying rural woodland if you are interested.

Why Buy a Woodland?

One of the common questions would-be woodland owners ask is, Can I make money by investing in woodland property? The cautionary answer is, likely not, certainly not enough in the short run to pay for the purchase of the woodland itself.

Many of the products available from woodlots do make money for a landowner, especially if you have invested in management over many years to bring your woodlot to a high state of productivity. But most rural landowners own their land for different purposes, either as a farm, as a residence, or both. Financial return from a woodlot is a bonus, and the initial financial outlay to purchase the land is not considered a cost to be balanced against the profit made from a timber harvest. You own the land anyway. Purchasing land with the hope that immediate profits from timber sales will pay for the purchase of the land is unrealistic.

The only example where financial returns can be significant over the long run is the case of a well-managed hardwood woodlot on highly productive soil. Hardwood trees, such as sugar maple, oak, cherry, and walnut, provide our most valuable timber; with excellent long-term management, such a woodlot can bring good financial returns. Little data is available to document profits and prices vary constantly, but we have heard anecdotal stories of ongoing financial returns as high as a farmer might earn growing corn.

Other types of woodlands, such as early successional stands, cedar or silver-maple swamps, and plantations, will generate little or no immediate profit. Only in the long run, with good management and a continued flow of products, will you make a profit; even then it is doubtful whether it would ever cover the cost of the land.

Most people buy land because they love it; they want the chance to walk through it, perhaps work to improve it in small ways, and often, to retire on it. You should buy land because you want to, for its own sake, then consider any financial return you can make from woodlot management as an extra benefit.

This is not to say you cannot make any profit. Unfortunately, it is all too common for landowners to harvest their woodland intensively just before selling it, in order to maximize their own returns. If you are lucky enough to find property where this is not the case, and the timber on the land is in good condition, there is certainly a good chance to make some immediate profit.

Steps in Purchasing Land

It is useful to make a few notes about your ideal property. Note what you want on your land. Do you want:

- a clearing where you can eventually build a house?
- year-round road access?
- an existing barn or shed?
- a stream or pond?

There are many different types of woodlands you can look for. On the other hand, location may be the most important factor for you. Price, of course, is a basic constraint. Make note of your ideas before you visit real-estate agents; you can easily get carried away by their enthusiasm, even if the properties they show you don't match your dreams. Having your desires written down may help to keep you on track.

If you have friends who already have country properties, talk to them about their experiences. What has worked, and what hasn't? Especially if you intend to live on the property, you want to be well aware of the advantages and disadvantages.

An initial visit to some real-estate agents and visits to some properties currently for sale will give you a sense of what is available and whether it comes close to matching your own dreams and pocketbook.

We advise visiting more than one real-estate agent initially, to find a person with whom you are comfortable and would like to deal . Then you may wish to ask for that agent's assistance in finding land for sale, by telling him exactly what you are after. You can also ask around the community; often older landowners may have land they are willing to sell to the right person, but have not yet listed for sale. Leaving your name and phone number with some local people might lead to some possibilities.

Investment in property is probably one of the biggest financial decisions you will make in your life, so take a planned approach and go slowly to be sure you are satisfied, especially if you intend to live on the property.

Property Investigations

If and when you find a woodland that matches your goals, there are several things you should check before making an offer to purchase. While you can make the resolution of these questions a condition of any offer to purchase, you want to deal with them quickly and carefully.

One is water. Even if you never intend to live on the land you buy, it is important to know if a well can be drilled there, and if you are likely to find a high-quality groundwater source for drinking. This can be answered by talking to the nearest well-drilling company, the local municipal engineer, and neighboring property owners. The cost of a well can be a major item, particularly if the land you buy is rocky.

The second related question is waste disposal. Any dwelling will have to have a septic tank and tile field-disposal system, or another option as approved by local and state or provincial regulations. This is perhaps the aspect of property ownership that is most regulated. Again, consultation with the municipality, local contractors who build septic systems, and with the relevant state or provincial agency will help you assess this. As with drilling wells, if your property is rocky or steeply sloping, installing a septic system may be a major expense.

Even if you do not intend to live on your land, you may want to build a shed or bunkhouse and an outhouse of some sort. Environmental regulations may prevent your doing this if you do not have an acceptable building site with a waste-disposal system. Therefore, it is important to assess these two factors carefully. Obviously, resale value may be heavily dependent on these factors as well.

A third thing to look at is road access. Existing road access may be seasonal or year round. If in doubt, don't believe what the real-estate agent tells you; check with the municipal office. If no road exists on the property, you may have to construct one in order to get access to the land you buy. Again, talk to local contractors to ask about the cost of doing this. As above, slopes and rocks can make building roads—even a simple tractor trail—expensive.

There are many other specific things you may want to check, such as the location of schools, the existence of stores, the extent of hunting, and so on. Many of these are more important if you are planning to live on the property. There are several other things you should ask about at the local municipal office in any case:

- What are the taxes and how fast have they been rising?
- What land-use bylaws would limit the uses you can make of the land?
- What other regulatory constraints may exist that will influence your land?

You also need to ascertain whether there is a survey of the land available. If not, you should be very leery about purchasing it without making this a requirement for the vendor, not yourself. We have all heard stories of purchasers buying their 10 acres (4 ha) of northern wilderness, then getting there to find that it is inaccessible and unsurveyed, so they are unable to even establish exactly which 10 acres (4 ha) of the woodland is theirs. In remote areas where adjoining property has not been surveyed, such work can be very expensive.

The last obvious thing to check on your property is the quality of the woodland itself. We have devoted two chapters in this book to preparing an inventory of your woodland. All of this will be useful to you in assessing the land and determining whether you feel it matches your dreams of being a woodland owner closely enough. Apply the inventory methods we have described to assess the woodland you are considering.

An alternative is to find a professional forester or wildlife biologist who will walk the land with you and give you his opinion. This is certainly a quicker way to initially assess the woodland. Don't let yourself buy land dreaming of maple-syrup production (or any other use) without assuring yourself that the woodland will actually support this.

If you get past all this and still feel the property you have found is right for you, then go ahead and buy it. Join the ranks of woodland owners and enjoy!

Further Reading

Boudreau, E. *Buying Country Land.* New York: Collier, 1973.

Vardaman, J.M. *How to Make Money Growing Trees.* New York: Wiley, 1989.

CHAPTER 12

ENSURING LONG-TERM CONSERVATION

Over the past three decades, the science of forestry has increasingly recognized the need to manage woodlands more sustainably. This has been reflected in international agreements growing out of a number of United Nations environmental programs that are grounded in concern for both environmental and economic sustainability.

The commitment of private landowners to woodland conservation occurs in the global context, and in eastern North America is the single most important contribution to long-term forest sustainability.

Although on a planetary scale most attention is given to the tropical rainforest, in the United States and Canada, forest conservation is also a serious concern. If the harvest of timber exceeds the natural regrowth in a region, the net result is a loss of economic stability in many small communities that depend on the forest industry. The loss of jobs in towns based on forestry puts a very human face on the issue of sustainability.

In addition to the economic impact, environmental or ecological benefits are being lost. Woodlands are critical to our water supply and can provide protection against soil erosion. Wildlife depends on woodlands for habitat, and we depend on woodlands for many forms of recreation. Woodlands make a vital contribution to the maintenance of biodiversity on the planet.

As a direct result of these concerns, industry and government are reconsidering the meaning of sustainability in forestry. The possibility that international trade in wood products would require some form of guarantee that all products have been produced sustainably – as is currently under discussion by the International Standards Organization – is a direct incentive to practice such measures.

Landowners, too, need to consider how their woodlands can be managed sustainably, whether their land will continue to supply them with an economic return through harvesting timber and firewood, or whether their property will contribute to conservation in a broader sense.

Changing Government Policies

While our main interest is in practical management options for environmental sustainability in woodland stewardship, it is worth noting that all the above concerns are not merely good words uttered by conservationists. Government policies are now changing substantially to reflect this new emphasis on sustainability.

One way to describe this is to say that our forest policies have evolved from single-purpose management, through a period of emphasis on integrated management, to what is now called ecosystem management.

For many years, forestry in North America was single purpose; just harvest the timber and forget about environmental impacts or long-term growth. Early in this century, conservationists recognized that such exploitation could not continue indefinitely. Then forest policies evolved to begin recognizing the need to integrate some concern for conservation and reforestation with harvesting practices, but timber harvesting has tended to still be the primary goal.

Today the term ecosystem management is used to reflect the need to consider the entire forest or woodland as an ecosystem. Some areas – such as land along waterways, wetlands, unique wildlife habitats, locations of rare species, or places of special cultural value – should not be used for timber production at all; other values should take priority.

This attempt to recognize all forest values and plan for woodland use in a more holistic way is the ecosystem approach. The report of the Ontario Forest Policy Panel provides one good example of evolving approaches. Suggesting that objectives for forest sustainability should include concern for biodiversity and natural-heritage forestland, as well as the quality of water, air, and soil, this panel recommended the following principles:

- maintain ecological processes essential for the functioning of the biosphere, and conserve biological diversity in the use of forest ecosystems
- use forest harvesting practices that emulate natural disturbances and landscape patterns
- do not harvest forest ecosystem types that cannot be returned to similar and healthy forests
- minimize effects on soil, water, remaining vegetation, wildlife habitat and other values
  — Ontario Forest Policy Panel, *Diversity: Forests, People, Communities*, 1993.

Whether at the state and provincial levels, or at the level of international trade, we are witnessing a significant new emphasis on environmental sustainability in woodland and forest management.

We turn now to some of the ways in which woodland owners can consider the long-term conservation of their land, and thereby contribute further to global forest sustainability.

Conserving and Caring for Your Woodland

The first and most important contribution you can make to woodland stewardship is simply to conserve your woodland, even if you do nothing else but keep your woodland intact.

Most landowners love their land, including their woodland, for its beauty and its wildlife, as well as the crops it can produce. If you do need to derive some economic return from your woodland, you can practice sustainable forestry with care, as discussed below. In fact, in the long run, sustainable forestry will bring more economic return than short-sighted exploitation.

Developing a carefully thought-out stewardship plan for your woodland, based on a good inventory, is probably the single most practical step you can take toward protecting your woodland in the long run. Compiling an inventory and developing your plan assists you in learning to understand and appreciate all the values of your woodland. By discovering perspectives you had not considered before, your own range of choices is widened.

A written stewardship plan is an excellent basis for discussing woodland management with your children or heirs. You cannot expect them to develop and carry on your love of the land unless you share your specific ideas with them. Even if selling your land, a written plan may be of great benefit to the next landowner. And some government support programs require a written stewardship or management plan.

You may wish to go beyond a stewardship plan and consider a legal agreement or donation that could guarantee longer-term conservation of your woodland (and possibly earn you significant tax benefits). As just one example, conservation easements are a legal technique that allow you to put specific conditions on the future use of your land, regardless of whether you sell it, leave it to your heirs, or donate it to a conservation organization. Under most conditions, you would also receive a significant tax benefit.

Many landowners are concerned about the cost of keeping their land and what will happen to it after their death. Woodlands that are not primary residences bring special tax concerns. There are many aspects to be considered in ensuring long-term conservation and many legal tools that may be of assistance to the landowner.

For all landowners, the cost of owning and managing a woodland is a concern. For most, consideration of your woodland as a component of your estate is relevant. And for a few, legal protection of your woodland in cooperation with a conservation organization may be of interest.

In this chapter, we introduce a number of concepts or tools that may assist you as a landowner in dealing with these issues. Because of widely varying individual circumstances, as well as legal differences between the U.S. and Canada, it is not possible to provide specific descriptions of these techniques. These options are presented for education purposes only; you are encouraged to seek your own legal advice.

Your Woodland as a Business

Some rural landowners own their woodland as a home away from home, a recreational property where their recreation consists of woodland management. Some owners have woodland on a property that is otherwise a farm or rural home. Some landowners depend on their woodland for a significant portion of their income. For many, paying the property taxes is an ongoing concern.

In many of these cases, treating your woodland as a business can mean a significant tax advantage. Even if you choose not to, understanding the financial positions involved may be illuminating.

Since tax law varies considerably, and is notoriously complex (and changes constantly), it is not possible here to do more than encourage you to consider this question. If you do have a reasonable expectation of profit from your woodland, even in the long run, it may be advantageous to treat it as a separate business, in which expenses can be deducted from income. The exercise of keeping careful track of expenses and taxes makes good business sense alone.

The Stewardship Plan

As we have emphasized throughout this book, your woodland-stewardship plan is the basis for your own commitment, as well as any of the following options. With its description of your woods, its documentation of your goals and objectives, and your preferred management options, your plan provides the basis for discussion of any future scenario.

The plan itself is your own first level of commitment. All too often, rural landowners never consider how they want to care for their woodlands. This leaves them vulnerable to the first unscrupulous logger who comes to the door offering to purchase their standing timber. All too often disaster results. Simply having a well-thought-out plan, with the understanding and commitment this implies, means you will make better, more informed decisions about your woodland. It is also likely you will gain substantially in your own appreciation of the woodland ecosystem.

Sharing with Others

The first level in expanding your commitment is to share it with others. In some cases this will mean your family. Children who are exposed to the beauty of woodlands when they are young will likely carry this interest for life. Heirs will understand your love of your woodland and your wishes for it, if you have shared your plans with them.

You may also gain by joining an organization that lets you meet others with similar interests. Whether it be a woodlot association, a natural-history group, or a recreational group (from hiking to hunting), sharing your own questions about your land will enrich your understanding. You may meet people who can assist you with information or tasks, or be able to help others yourself. Often people find that they first realize how much they know about their own land and their commitment to it upon sharing this with others.

Voluntary Agreements

A number of conservation organizations have developed voluntary stewardship agreements, which serve to encourage a commitment to conservation by private landowners. Known variously as landowner-contact, landowner-notification, or private-stewardship programs, these offer the opportunity for a landowner to be part of a larger conservation vision. Most are strictly voluntary and educational. They may serve to provide contacts and information through which you can get further assistance in managing your own land, or they may help you meet other like-minded landowners nearby. Such an agreement will help you to realize and understand the significance of your own woodland in the context of others in your region.

As a footnote, the writing of this book is in fact a result of the authors' involvement in such stewardship programs. Through involvement in landowner-contact programs over several years, we have found many landowners who want to understand more about managing their woodlands, but aren't sure where to start. We started writing this book, in part, to provide some answers for those interested.

Management Agreements

A management agreement is usually a written agreement between a landowner and a conservation organization that goes one step beyond a voluntary or a handshake agreement. An owner may, for example, agree to give a fishing club access to a stream, in return for help from them in building a fence. Other owners may lease their woodlands to hunters in return for help in controlling trespassers. Usually such agreements are best put in writing, rather than left as verbal agreements, so parties to the agreement know their responsibilities.

Often conservation organizations or government agencies that provide grants or tax incentives to landowners will also require a management agreement in some form. If you have prepared your stewardship plan carefully, it should provide all the information needed. Usually when an agency or organization is giving out funding to landowners, a written agreement is required; usually a time limit is specified.

Long-Term Legal Protection

There are several longer-term legal tools that can assist landowners who wish to ensure the appropriate use of their woodlands by heirs or future owners. Not only do the relevant laws differ between the U.S. and Canada, but they vary to some extent by state and province as well, so you are well advised to consult an organization in your own region, as well as your personal lawyer.

As examples, we list the following four techniques for longer-term legal protection:

1. Conservation Easement: This is a legal agreement that should last in perpetuity. It can specify any number of requirements with which future landowners must comply. Conservation easements are usually donated or sold to nonprofit conservation organizations, such as land trusts, and the landowner may gain a tax benefit as a result, assuming the conservation easement restricts the uses that can be made of the land, and therefore lessens its value.

   Such easements are often used in creative arrangements that permit some development in return for protection of other features. Usually they are used to prevent action, such as the destruction of woodland; it is more difficult to enforce active management.

2. Restrictive Covenant: This technique is similar to a conservation easement, but simply places a restriction on allowed uses of the land for a period of time. It may be negotiated between landowners or between

a landowner and an organization. It has fewer tax benefits and does not last in perpetuity.

3. Family Trust: Family trusts are legal arrangements that enable shared or arm's-length ownership of assets, such as woodland, among members of a family. Typically, they are established by an owner of an asset if the owner wishes to share the benefits of the asset as an ongoing operation, without requiring family members to get involved in direct management and without selling the asset. Often an arm's-length trustee is named, in which case family members are usually named as beneficiaries. Such arrangements are only limited by trust law (which is complex) and the creativity of the grantor.

   Family trusts should be distinguished from land trusts, which are nonprofit charitable organizations usually devoted to the conservation of land in a specific region.

4. Profit à Prendre: This arrangement is usually a written agreement between a landowner and a second party that enables the second party to profit from something, such as timber on the landowner's land. It is attached to the land title and, therefore, extends to future owners of the land.

Donations and Bequests

In some cases, a landowner may consider giving his land to a conservation organization, either while he is still alive or as a bequest in his will. There are numerous variations that are possible if this is of interest to you.

- Donation: a gift of land, usually with direct tax benefits, though calculation of these may be complex, especially if the property is not your primary residence and is subject to capital-gains tax.

- Bequest or devise: a donation made at the time of death through stipulations in an owner's will.
- Conditional donation or bequest: landowners may also stipulate conditions attached to either a donation or bequest, such as continued conservation of the land being donated. Stipulating conditions in a bequest is a risk, since the organization receiving the bequest may not accept the conditions, but after death it is not possible to negotiate details. Landowners are advised to approach organizations and discuss their intent, rather than simply leaving the gift in their will.
- Bargain sale: a bargain sale is a sale at less than market value or, in other words, part sale and part donation. It may be ideal for landowners who require some financial return, but are willing to donate some of the value of their land.
- Life estate: this enables a landowner to donate his land while retaining the right to live on it for the remainder of his life.

Planning Your Estate

With all these legal tools available, surprisingly few landowners make use of them. In many cases, there are significant financial advantages to be gained by using such tools, although details will depend on an individual's personal financial circumstances. Owners who wish to see their own woodlands cared for in the distant future would be well advised to consider them.

It is undoubtedly true that woodland landowners would generally benefit by giving greater thought to the planning of their estates, whether it be as simple as discussing their hopes with heirs or a complex legal arrangement to protect their woodlands in the future.

Further Reading

Haney, H.L., and W.C. Siegel. *Estate Planning for Forest Landowners.* New Orleans, LA: Forest Service, Southern Forest Experimental Station, 1993.

Hoose, Phil. *Building an Ark: Tools for the Preservation of Natural Diversity.* Covello, CA: Island Press, 1981.

Landowner Resource Centre. *Making Your Woodland Pay: Financial Aspects of Property Management.* Manotick, ON: Landowner Resource Centre, 1997.

Milne, J. The Landowner's Options: *A Guide to the Voluntary Protection of Land in Maine.* Bangor, ME: Maine Critical Areas Program, 1985.

Mogan, P.J., and G. Nemeth. *Family Trusts for Tax and Estate Planning.* North Vancouver, BC: Self-Counsel Press, 1996.

CHAPTER 13

The Spirit of the Woods

L andowners may own woodlands for many different reasons, and may manage them for many different uses. Some find pleasure in watching the spring wildflowers emerge and the migrating birds return. Others enjoy walking with grandchildren when the fall colors are at their peak. Some spend their time in woodlands in March and April, making maple syrup. Others enjoy the challenge of managing their woodlands for sustainable timber harvesting.

Whatever their purposes or interests, virtually all woodland owners have experienced the spirit of the woods. There is something special about being in the woods, absorbing the natural world around you. This is especially true if you can be there often enough to feel you belong, to feel you know the woodland all year round. Ultimately, it is this connection with the earth that makes woodland owners feel they have a special relationship with a place, a woodland to which they can return again and again throughout their lives.

We hope this book has opened your eyes to a greater understanding of the woodland ecosystem in all its complexity. With greater understanding comes greater appreciation. Woodlands are a wonderful part of nature. Manage yours with care. We leave you with our own observations of the seasons in the woods.

Seasons in the Woods

Winter

The cycle of seasons in the woods begins in the deep of winter, when the ground is covered with a foot or more of snow and all is still. Trees, shrubs, and ground cover are all dormant, and thousands of living creatures are either hibernating or at a dormant stage in their life cycles. The woods are quiet and seem almost deserted. Most birds left for warmer climates in late summer and early fall. Turtles and frogs are asleep in the mud at the bottom of the wetland, bats are hibernating in hollow trees, and mice are asleep in their nests. Larger mammals, such as the porcupine and muskrat, have retreated to their borrows for hibernation. Millions of insects are hidden in the litter on the forest floor, or under bark, as eggs or larva, waiting to emerge in the spring.

After a heavy snow even the active wildlife retreats to stillness. Snow decorates the thinnest branches, and the woods looks like a winter wonderland. But after only a few hours it comes to life again. In our woods, the most common tracks are those of the squirrels, especially in the corners where groves of hemlock grow. Along the edge of the field where brush piles give shelter from coyotes, a few rabbit tracks can be seen. There are deer trails out of the cedar swamp through the upland woods and across the field, where deer have wandered in search of tender shoots to nibble on. Woodmice are awake and busy, traveling through their hollow tunnels under the snow.

Snow conditions make an enormous difference in your ability to identify tracks easily.

- Deep powdery snow reveals only fuzzy tracks, or the loose holes left by footprints.

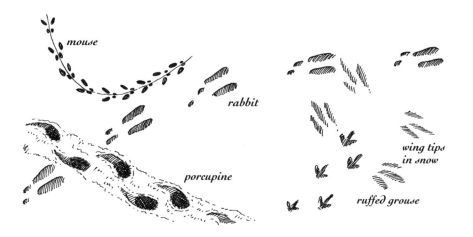

- A typical 2- or 3-inch (5–7 cm) snowfall enables fairly clear tracks, especially if the weather is mild.
- A hard crust or ice cover on the snow may mean there are no tracks at all, only slight scrapes on the surface.
- The best conditions for clear wildlife tracks are a very light snowfall on top of a hard crust. Then even the tracks of tiny mammals, such as woodmice, stand out clearly.

Within days after a snowfall, especially in mild weather, the ground in some areas of the woods will be so crisscrossed with tracks that you can no longer follow individual trails. But over a season, you begin to recognize the trails that wildlife uses repeatedly. A red squirrel travels from the hemlock grove 300 feet (90 m) back in the woods along the fence line, sometimes from tree to tree, until it reaches our old cedar split-rail fence. The fence is its highway to our feeder, where it robs the birds of a small share of the seed. The white-tailed deer come out of the cedar swamp at the far side of the woods and walk on a diagonal trail heading northwest. Skirting the low area, they always cross the wetlands at the narrowest point and head back along the fence line to the west. Day after day, their tracks indicate a nocturnal visit, always following approximately the same route.

The birds that brave our winter temperatures soon make the woods sound alive, too. Blue jays scream in the cold winter air;

chickadees chatter to each other as they search for seeds in the hemlock trees. A white-breasted nuthatch calls its nasal "hank, hank" from the trunk of a maple. Occasionally a downy woodpecker's tree-tapping hammer can be heard echoing through the woods. In our yard at the edge of the woods, mourning doves, tree sparrows, and juncos brave winter to come for food. A bright red cardinal appears for a time in early morning, and sometimes the aggressive blue jays appear, momentarily driving the other birds to shelter. Only for a few hours, after a deep snowfall in the deepest cold, does the woods actually sleep.

Maple-Syrup Season

Snow is still on the ground in March, when maple-syrup season begins. Gradually it melts, often forming icy patches on the trail. Bare ground appears between the deeper snow patches, and the top inch of soil thaws. But the ground is still frozen an inch or two down, so if you step on the soil you leave muddy tracks.

In the woods, hibernating wildlife is waking up. The tracks of the porcupine, from tree to tree and across the road to the barn, reveal its nighttime wanderings. Squirrels arrive at our woodside feeder in record numbers for these few short weeks before the snow melts entirely. Deer tracks are more frequent.

For a month now, the great horned owl has been calling occasionally, and the red-tailed hawk screams as it drifts over the field. The robins have returned, searching for grubs and worms in small patches where the snow has melted. Red-winged blackbirds call from the edge of the swamp. The first wave of migrating birds is back, to join those that have stayed the winter. A large flock of Canada geese, which often resides throughout the winter where the river stays open, have returned to their pairs for nesting. You hear them honking in the distance, then see them flying overhead, sometimes so low you can clearly hear their wing beats. Pairs of geese land on ponds still frozen with ice, looking for all the world as if they can't understand where the water went. Only rarely, a

spectacular flight of swans passes overhead, barking their calls on their way north.

Thousands of families lucky enough to own a woodlot know this time of year as maple-syrup season. From late February until late April, depending on your location, trees are tapped, and maple sap is boiled down to make syrup. The days are milder now and often sunny. These are beautiful days to be working in the woods, a refreshing time to be out of doors after the long winter.

Maple trees are unique in the woods, coming to life with the movement of water and nutrients through millions of tiny channels up and down their tree trunks, several weeks before other tree species. Scientists still do not completely understand the mechanism of the maple-sap run, but it is not simply sap moving from the roots to the branches to be ready for the emergence of leaves in the spring. During sunny days in the early spring, when temperatures rise above freezing, the fine twigs thaw, then freeze again with nightfall. This creates frost inside the wood fibers and a buildup of osmotic pressure, which draws sap upward in the trunk and branches during the cold night.

As temperatures rise above freezing in the morning, the fluid thaws and, partly by gravity, it drains downwards (and out if the tree has been tapped) as sap. Repeated cold nights and sunny days create the best sap runs. In fact, the sap run will usually proceed in fits and starts according to the weather.

With luck, even the amateur syrup maker can produce very high-quality syrup, though this may not be recognizable to the average urban consumer, familiar only with the thick, dark syrup commonly sold in stores. The best early run of maple sap produces a syrup very light in color and taste, the prize-winning stuff of many a woodlot story.

The Spring Melt

April is the season of melting snow and of water. As it begins, the ground is still frozen, and melting snow runs off quickly into rivulets, streams, and wetlands. Sometimes, on a particularly warm day, the snow seems to vanish into thin air, evaporating directly.

But as the month proceeds, the ground thaws, first becoming a thin layer of saturated soil over still-frozen ground, then a giant sponge which water infiltrates. Draining across the surface and through the ground, water seeks a downhill course. Temporary streams flow in all the tiny valleys. On the sides of all the slopes, springs form, as water seeks its way downwards in the soil, encounters a layer of soil that is harder to infiltrate, and then seeps sideways to emerge again on the surface as a spring.

This is the hydrological system – the drainage of water on the surface and as groundwater – that feeds the plant growth soon to come. Never is this hydrological system more obvious than during this brief period of the spring melt. Shortly, the surface of the land will dry out, and water will flow only in stream channels or rivers, and accumulate only in ponds and wetlands.

But for a few short weeks the entire woods is a waterway, as the snow melts and groundwater is replenished. The surface of the forest floor is awash as spring comes to the land.

When spring is full-blown, the melt is often sudden and fast Ice on the pond changes from safe to thin in a warm week; then it gradually melts, first around the edge, and turns to slush. On a final warm day, the ice of the entire pond can disappear in twenty-four hours, as it crosses the threshold from freezing to unfreezing.

This is the time to walk your woods and watch for the environmentally sensitive seepage areas. Only wet on the surface for a few weeks of the year, they are vital links in the drainage pattern and should always be protected from disturbance.

Spring melt in the woods also reminds us of the rapid runoff that occurs on the rest of the landscape where forest cover has been removed. Here there is little slow seepage; water runs rapidly into streams and rivers in the worst circumstances, causing disastrous flooding. Protection of forest cover is perhaps the single best way to protect the hydrological system.

Even as the snow melts, the first plants emerge and bloom. The lowly skunk cabbage is one of the first, poking its hooded spike of reddish green out of the swamp floor, while still surrounded by melting ice and snow. A small flower blooms within the hood before most plants have even emerged.

Later, the bright green leaves of the leek or wild onion emerge in patches in the woods. At the same time, the ghostly purple stem of the blue cohosh rises out of the forest floor, on warm days growing over 1 inch (2.5 cm) a day. Soon it will unfold, gradually turning green and looking like an ordinary part of the woodland ground cover.

The earliest birds – juncos, mourning doves, and house finches – are already nesting, visible in pairs as they mate and construct

their nests. The owl may already have young it is feeding deep in the forest.

The second wave of spring migrants has begun to appear, with kinglets, flickers, and meadowlarks returning or passing through on their way north. Great blue herons can be seen flying over in the evening on their way to roost, and turkey vultures soar over the fields and woods, holding their wings in a shallow V-shape.

Spring begins in earnest with the first of the frog choruses, the gabbling duck-like calls of the wood frog. When heard for the first time, it would seem that a large flock of ducks has just landed in the wetland, but you'd find not a single bird there, instead the water is moving with hundreds of frogs.

Shortly after, the spring peepers join the chorus. Over the next eight weeks you can follow a clear sequence of frog calls as different species in turn join the chorus with their unique cadence until mid-June, when the deep-voiced bullfrog signals the end of the amphibian mating season.

At the same time, the buds on trees in the woods are swelling, and spring wildflowers are poking above the forest floor, ready for the rush of real spring as the weather warms.

Migrating Birds and Spring Wildflowers

As the wet spring season dries out and real warmth returns, the most beautiful three weeks of the year in the woods begin. During this short period, the main wave of migrating birds, especially warblers, passes through. The woods is alive with bird songs. Dozens of species of spring wildflowers bloom, getting in their short periods of growth before the leaves on the trees close off the direct sunlight.

This is undoubtedly the most wonderful time to walk in your woods. With a little patience and a field guide, you can learn to identify the spring wildflowers. With a pair of binoculars you can catch glimpses of the birds in the treetops. The incredible diversity of life in your woods comes alive before your very eyes. And the mosquitoes are not out yet, making a walk in the woods a pleasant experience for the last time until mid or late summer.

The magic ingredients in this awakening of the woods are temperature and sunlight. Penetrating through to the forest floor in the early spring, sunlight brings out the wildflowers: wild leek, spring beauty, hepatica, red and white trilliums, bloodroot, wild ginger, jack-in-the-pulpit, dutchman's-breeches, the ghostly blue cohosh, and others. In wet seepage areas, the coltsfoot lifts its yellow flower, somewhat similar to the common dandelion on your front lawn, and down in the wet swamp, the large green leaves of the marsh marigold are rapidly followed by a carpet of bright yellow blooms. Drawing their energy from the sunlight while they still can, these plants emerge, grow, and flower all during this short period of early warmth.

Another wave of migrating birds returns, many of these passing through on their way north to the boreal forest, where they will nest. These species, mostly warblers, must wait until a measure of warmth appears in that northern forest before migrating through our region.

Passing through our own woodland are yellow and yellow-rumped warblers, the black and white, the blackburnian, chestnut-sided warblers, as well as the redstart. Birds that will stay in the woods and nest may include the increasingly rare wood thrush, as well as the veery, the oriole, and the ovenbird. Along the edge of the woods, the chipping sparrow calls its steady chip, and the

white-throated sparrow, the song sparrow, and the field sparrow begin to nest. Over the open fields, the swallows have returned.

Meanwhile, the chorus of frogs continues to change. The spring chorus frog soon joins the spring peepers, then the guttural-sounding leopard frog joins in. By the time the leaves are out, the trill of the American toad can be heard regularly, then the shorter barking trill of the tree frog takes over the night.

Almost every evening the female porcupine wanders out of the woods into the hay field to nibble on nutritious shoots of young afalfa.

Toward the end of this magical period in the woods, an incredible transformation takes place as leaves burst from their buds, growing to nearly full size in a matter of days. At perhaps a quarter-million leaves per tree, or billions in a woodlot, this growth, based on enormous reservoirs of stored nutrients, water, and energy, is truly one of the miracles of nature.

Nesting Birds and Mosquitoes

Soon after the leaves emerge, the woods becomes inhospitable – at least to people – as mosquitoes hatch by the millions, often joined by blackflies. But these bugs, while the bane of humans, are a flying banquet for insect-eating birds.

Birds are now nesting at all levels in the woods, from those that build their nests directly on the ground, such as the ovenbird, to those that prefer the forest canopy, such as the oriole.

Nest locations are directly tied to the layers of the forest structure. Those on the ground are most susceptible to destruction, especially by wandering domestic cats and dogs. This is the best time to stay out of the woods, leaving it as a haven for birds and bugs, at least for a month or two. It is a very important time to control cats and dogs; domestic cats are estimated to be the number-one predator of songbirds.

In the pond, the deep note of the bullfrog marks the end of the spring mating season for amphibians. And the turtles, which emerged earlier to sit on logs soaking up the warmth of the sun, are now seeking sandy soil in which to lay their eggs. They will then disappear into the water again, rarely seen again before the following spring.

By early June, the leaves on the trees are shading the forest floor, and the wildflower season is over. But the main period of growth in the forest is now underway. Though totally invisible to our eyes, trees are steadily growing outward in diameter and upward in height. New cells in the cambium layer just inside the bark divide to become the new wood in the tree.

Trees extend their branches and shoots at the outer edge, expanding their canopy of leaves and growing taller. All species have their own regular growth patterns, with buds and new twigs either emerging exactly opposite one another or alternately down the stem. Such characteristics can help you learn to identify trees even without their leaves (as you must if you wish to make maple syrup).

Trees also have their own timetable. The white ash and basswood burst into bud and leaf considerably later than do beech and sugar maple, leaving them looking for a week or two as if they might have died over the winter, until you see the emerging flush of green.

In the spring, the new growth occurs quite quickly, and the newly created cells in the cambium are large, but by fall growth has

slowed markedly, and the new cells are quite narrow in structure. This difference provides a line of lighter colored and darker colored cells that look like rings on the stump when a tree is cut, thus giving the annual growth rings you can count to determine the age of a tree.

Out on the edge of the forest, this is the season when young tree seedlings have to compete with rapidly growing grass. While the pines and spruce we use for reforestation can compete with the dense grasses of an old field, deciduous trees, such as maples, cannot. They start from germinating seeds in the early spring, but get overwhelmed by the other plants and stall in their growth, most of them staying forever as small seedlings or saplings.

Late Summer and Early Fall

By mid to late summer, growth in the woods has slowed and birds have finished nesting. Many birds have begun to migrate south again, some as early as mid-August. The young of mammals have grown to full mobility, if not independence. Eggs left in the pond by mating frogs and toads in the spring have hatched; tadpoles have grown into frogs, and some species have left the pond in search of other habitat.

Outside the woods, the harvest is in full swing. Even in early August, farmers are combining wheat in the fields, and a second cut of hay is headed for the barn. Late summer is the season of fresh tomatoes and sweet corn, and the best time of year to visit a local farmers' market.

As the mosquito population declines, walking in the woods becomes tolerable again, and after the first frost, a walk is downright pleasant.

With the arrival of fall, the best season for working in the woods returns. September is a great time to learn to identify trees if you are planning to make maple syrup or do some forest-management work. Leaves are still on the trees, but the bugs are gone, so you can spend time in the woods easily with your field guide.

On the edge of the woods or in the open meadow you can see a measure of this year's growth. Saplings may be a foot taller. Pines planted for reforestation put on one new whorl of branches each year, so you can look at the last tall new shoot to see the measure of

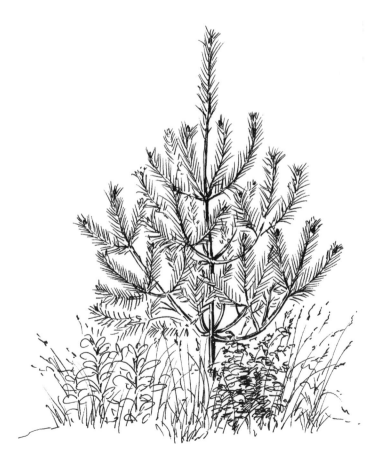

their growth, and you can count branch whorls to tell the age of the tree.

Though mosquitoes decline in the woods, insect life is in full flight elsewhere, with butterflies, grasshoppers, and cicadas filling the meadows or calling from trees. The jewelweed reaches maturity as its seed pods stretch; they explode at a touch, giving it the common name of touch-me-not. Queen Anne's lace turns roadsides white with its blooms, while old fields turn yellow and purple with the color of goldenrod and asters.

Like early spring, early fall is a great time to introduce children to the beauty of your woodland. With no bugs to bother you, you can spend time searching for frogs or learning about plants. The wide-eyed sense of wonder children experience in seeing the woods up close reminds us adults of the beauty we tend to take for granted.

Fall Colors

Fall colors begin as early as late August in some northern areas, but the peak few weeks of beautiful color usually occurs in October. Bathed in orange, reds, and yellows, the woods is transformed again into another world. A walk in the woods on a sunny day at the peak of the fall colors is a magical experience indeed.

Some of the colors have been there in the leaves all season, but they are masked by the intense green caused by the chlorophyll in the leaf. When temperatures fall, photosynthesis stops, chlorophyll production stops, and the intense green fades, leaving the yellow and orange to show through. These are the same colors found in carrots and squash. If the weather is particularly cool and sunny, more red tends to develop in the leaves in relation to the acidity of the sap.

In any case, the effect is astounding. In a few short weeks, the woods changes from green to yellows, orange, and red, then the leaves fall to the ground, recycling the enormous growth of the summer season.

While maples in the woods add the brightest colors, other species follow suit. Trembling aspen paint the hills gold, as do tamarack a little later in the season. Even white pine contribute as the inner needles, now two years old, turn gold and fall.

Frost, as well as dwindling daylight, is a magic ingredient in this seasonal change, and it does more than bring on fall colors. The last of the mosquito population dies, opening the possibility of woodland walks again. Mice and other creatures feel the cold and seek to share the warmth of your house over the winter. It's a battle to keep them where they belong.

Cones, nuts, and seeds ripen. Squirrels, chipmunks, and other wildlife harvest the bounty, storing energy, and food itself, for the winter to come. The last of the birds are headed south, and mammals are preparing to hibernate. The noisy frogs and toads of spring have found a place to burrow in the mud until next year.

Deer Rutting and Firewood Cutting

Working in the woods is pleasurable and most productive in late fall and early winter. This is the time to cut firewood so it can dry a season before next year's maple-syrup production or for use in the wood stove. You will recognize the truth of the saying "firewood will warm you twice" after spending a few hours cutting and stacking it.

Using the leftover smaller branches, you can build a brush pile for wildlife to use as a shelter. Place a few large branches at the bottom to ensure some large cavities, then cover the base with smaller branches. This will provide a secure shelter for a fleeing rabbit.

The woods takes on a different atmosphere at this time of year, the bare trees looking like gray skeletons in the moonlight. With the leaves gone, you can clearly see the structure of the trees and watch the swaying of the tree trunks in a strong wind. Nights are clear, and the stars seem to shine brighter in the late fall, making it easier to pick out the constellations.

Many birds have migrated; those that stay for the winter will benefit from a handout. Putting up a bird feeder to attract birds over the winter will enliven your yard considerably.

We meet our neighbor out hunting in November. Though not hunters ourselves, we have a mutual agreement. He hunts on our land as well as his own, and helps me deal with trespassers. In turn, I dine on venison steak once or twice over the winter.

On walks we notice bare patches of earth where deer have kicked away the leaves and scraped the soil during their fall rut, or mating season. Deer are more active at this time of year, wandering further afield than they will in the dead of winter. The early snow-falls of November and December will enable you to follow their trails through the woods. Small mammals are also active, storing the nuts and fruit that will see them through the cold. It is this preparation for deep winter that signals the end of the cycle of seasons.

Further Reading

Most of the seasonal changes described in this chapter are based on the authors' own personal observations. Similar accounts of changing seasons include

Beston, Henry. *The Outermost House*. New York: Ballantine, 1928.

———. *Northern Farm*. New York: Ballantine, 1928.

Brooks, Paul. *Speaking for Nature*. San Francisco: Sierra Club, 1980.

Dillard, Annie. *Pilgrim at Tinker Creek*. New York: Harper and Row, 1974.

De Kiriline Lawrence, Louise. *To Whom the Wilderness Speaks*. Toronto: McGraw-Hill Ryerson, 1980.

Leopold, Aldo. *A Sand County Almanac*. New York: Ballantine, 1949.

Mitchell, John H. *A Field Guide to Your Own Back Yard*. New York: Norton, 1985.

GETTING HELP WHEN YOU NEED IT

One of the most useful things you can do as you learn how to manage your woodlot is to have a helpful advisor who has the experience to give you practical advice, or confirm that you are headed in the right direction.

Finding a knowledgeable professional or experienced local landowner can be one of the most encouraging things you can do to further your own knowledge and confidence. Joining local, provincial, or state organizations of other landowners sharing similar interests is another positive aspect of forest ownership. By sharing ideas, asking questions, attending workshops, and getting others to visit your land, you can come to understand your own woodlot and your management options much better.

Throughout the book we have indicated a number of topics on which you might want to seek an expert opinion. We encourage you to take advantage of every opportunity.

Professionals come in basically two types. In most states and provinces, there is government extension staff available to provide free advice to landowners. In the U.S., these experts are often associated with a university-based extension service. They have appropriate professional training and usually have wide experience.

Often there will be a team of experts available behind the scenes that they can call on, in turn, to provide detailed advice on specific topics. For some services, there may be a specific fee charged.

On the other hand, there are also many independent professional forestry consultants available; in some jurisdictions they are taking over the role of former government extension staff, lost due to budget cutbacks. These consultants also have appropriate professional training and usually have many other contacts to whom they can direct you for specific advice. Usually they will charge a fee for their time, but they should always be prepared to come and meet you, answer a few questions, and talk about your needs before you actually hire them.

Both types of professionals can direct you to other experts on particular topics, other information sources, and to relevant organizations. They can also help you learn about any financial-support programs available. It is through these individuals that you can get the names and addresses of specialized organizations, such as Christmas-tree growers or maple-syrup producers. You can also find sources of equipment and trees themselves.

Most state or provincial extension agencies have numerous publications and often have videos available, usually at low prices. Since these are written for more specific jurisdictions, they are very useful in providing details of available programs and support. Start your own library of useful information.

We encourage you to build up your own local list of advisors upon whom you can call for advice. Over time you will find that the effort put into this is one of the most supportive things you can do to help you with woodlot management.

Working with Professionals

The professionals you may call on, either government staff or private consultants, are independent experts you can count on to provide an objective look at your woodlot, providing you tell them what you want. They may offer a complete range of services, from just a little advice over the phone, to the complete development of a woodlot-management plan and the carrying out of a marked-timber harvest.

Professionals should be willing to tell you clearly what services they offer and what their fees are, including how payment should be made. You should feel free to consult several before hiring one to work for you. They should also be willing to tell you when they do not have the expertise to answer your questions. Many will be foresters, trained to manage timber harvests; if your questions are about wildlife, they may not be able to help directly. If they appear to gloss over your questions and redirect your attention to the topics they know best, you may want to hire someone else.

Remember to discuss your interests thoroughly; professionals cannot serve you well unless they understand what you really want. Agree in writing as to exactly what a consultant will do and the form of any report to you. You want to be able to understand their work. The consultant you hire can provide a basic contract that you can modify and both agree to. Professionals should also be willing to give you references of others for whom they have worked.

Conflicts of interest can arise when a logging company provides the consultant to mark your woodlot for harvest, or worse yet, when you skip the consultant and let the logger mark the woodlot. There are numerous stories of log buyers sitting in farmhouse kitchens waving fistfuls of thousand-dollar bills at landowners, offering what seem like high prices to cut their woodlots. In far too many cases, the landowner discovers later that his woodlot has been devastated. In almost every case, the landowner would have been far better off financially to hire an independent consultant to evaluate his woodlot and supervise the timber harvest. So start with a professional in the first place, and you will benefit in the long run.

Landowner Organizations and Opportunities

Joining an organization, such as a local woodlot-owners' group, or a state or provincial forestry association, can bring you into contact with many other landowners like yourself and bring you useful information. These organizations are another source of addresses and contacts for more specific needs. They usually have newsletters, meetings, and often run workshops or tours which you can take advantage of.

In some states and provinces, there are volunteer training programs available to landowners. Programs such as the Master Forest Owner Program, the Community Woodland Steward Program, or the Coverts Program (which emphasizes wildlife habitats) provide training for interested landowners, who then volunteer to speak to others if requested. Getting involved directly with other landowners in programs such as these can be very enriching. Use professional contacts to find out whether such opportunities exist in your area.

Financial-Support Programs

Many government agencies, as well as nongovernment groups, provide financial grants or loans for specific purposes. These are so numerous that a list for even one state might run to several pages; a recent Minnesota brochure lists thirty-seven different opportunities for landowners.

Financial grants may be available to help develop woodlot-management plans, to help with the cost of management activities, and to fund tree planting, streambank rehabilitation, and many other related activities. In all cases, there are likely specific rules and criteria that must be met before you can qualify; most such programs are very specific.

It is, therefore, necessary to inquire within your own state or province in order to find out what is available. It is well worth the effort; some financial-support programs can run into thousands of dollars for individual landowners. In many cases, receiving financial support will require that you have a woodlot-management plan and are actively managing your woodlot according to the plan.

Tax Programs

There are numerous tax programs related to woodlot ownership. These range from the tax rules that govern businesses—some of which can apply to woodlot operations—to the taxes governing estates and inheritance. A number of jurisdictions have tax-incentive programs of one sort or another, often involving a reduction in property tax in return for specific efforts at woodlot management.

Because tax issues are complex, occurring on many levels and varying with each jurisdiction, there is no alternative but to inquire as to the rules in your own state or province. Although this may become tedious (tax rules are the domain of bureaucrats after all), the financial benefits can be significant. A 75 percent reduction in property tax, as the Ontario Managed Forest Tax Incentive Program provides, repeated year after year, can save a landowner many thousands of dollars over a lifetime. It is well worth taking the time to investigate opportunities that may apply to you.

The references that follow can provide further reading on specific topics; the addresses should provide you with initial contacts within your region. Together, they should enable you to access the help you need to carry out your woodlot management.

The authors welcome correspondence from readers of this book at:

The Centre for Land and Water Stewardship
The University of Guelph
Guelph, Ontario N1G 2W1
Canada

Further Reading

GENERAL

Beattie, Mollie, Charles Thompson, and Lynn Levine. *Working with Your Woodland: A Landowner's Guide.* Hanover, NH: University of New England Press, 1993.

Brett, R.M. *The Country Journal Woodlot Primer.* Brattleboro, VT: Country Journal Publications, 1983.

Fazio, James R. *The Woodland Steward.* Moscow, ID: The Woodland Press, 1985.

Minckler, L.S. *Woodland Ecology: Environmental Forestry for the Small Owner.* Syracuse, NY: Syracuse University Press, 1980.

Walker, Laurence C. *Farming the Small Forest: A Guide for the Landowner.* San Francisco: Miller Freeman, 1988.

Wiskel, Bruno. *Woodlot Management.* Edmonton, AB: Lone Pine, 1995.

ECOLOGY

Hunter, M.L. *Wildlife, Forests, and Forestry: Principles of Managing Forests for Biological Diversity.* New York: Prentice Hall, 1980.

Landowner Resource Centre. "The Old-growth Forests of Southern Ontario" (fact sheet). Manotick, ON: Landowner Resource Centre, 1996.

———. "Restoring Old-growth Features to Managed Forests in Southern Ontario" (fact sheet). Manotick, ON: Landowner Resource Centre, 1996.

Richberger, W.E., and R.A. Howard. *Understanding Forest Ecosystems.* Ithaca, NY: Cornell University Extension Service, 1980.

Yahner, R.H. *Eastern Deciduous Forest: Ecology and Wildlife Conservation.* Minneapolis: University of Minnesota Press, 1995.

WILDLIFE

Cox, Jeff. *Landscaping with Nature.* Emmaus, PA: Rodale Press, 1991.

Decker, Daniel J., and John W. Kelley. *Enhancement of Wildlife Habitat on Private Lands.* Ithaca, NY: Cornell University Extension Service, 1986.

Decker, D.J., et al. *Wildlife and Timber from Private Lands: A Landowner's Guide to Planning.* Ithaca, NY: Cornell University Extension Service, 1983.

Hassinger, Jerry, et al. *Woodlands and Wildlife.* University Park, PA: Pennsylvania State University, 1979.

Henderson, C.L. *Landscaping for Wildlife.* St. Paul, MN: Minnesota Department of Natural Resources, 1986.

Hunter, Malcolm L. *Wildlife, Forests and Forestry: Principles of Managing Forests for Biological Diversity.* New York: Prentice Hall, 1990.

TIMBER

Goff, Gary R., James P. Lassoie, and Katherine M. Layer. *Timber Management for Small Woodlands.* Ithaca, NY: Cornell University Extension Service, 1994.

McEvoy, T.J. *Introduction to Forest Ecology and Silviculture.* Burlington, VT: University of Vermont, 1995.

Staley, R.N. *Wood: Take a Stand and Make It Better.* Toronto: Queen's Printer, 1990.

Van Ryn, Debbie M., and James P. Lassoie. *Managing Small Woodland for Firewood.* Ithaca, NY: Cornell University Extension Service, 1987.

REFORESTATION AND NATURALIZATION

Daigle, J.M., and D. Havinga. *Restoring Nature's Place.* Schomberg, ON: Ecological Outlook Consulting, 1996.

Henderson, C.L. *Landscaping for Wildlife.* Minneapolis: Minnesota Department of Natural Resources, 1987.

Johnson, Lorraine. *The Ontario Naturalized Garden.* Toronto: Whitecap, 1995.

Landowner Resource Centre. "Careful Handling and Planting of Nursery Stock" (fact sheet). Manotick, ON: Landowner Resource Centre, 1995

———. "Clearing the Way: Preparing the Site for Tree Planting" (fact sheet). Manotick, ON: Landowner Resource Centre, 1995.

———. "Cover Crops Help Tree Seedlings Beat Weed Competition" (fact sheet). Manotick, ON: Landowner Resource Centre, 1994 .

Martin, L.C. *The Wildflower Meadow Book: A Gardener's Guide.* Charlotte, NC: Faast and McMillan, 1986.

Ministry of Natural Resources. *Managing Red Pine Plantations.* Toronto: Ontario Ministry of Natural Resources, 1986.

TreePeople. *The Simple Act of Planting a Tree.* Los Angeles: J.P. Tarcher, 1990.

Weiner, Michael A. *Plant a Tree.* New York: Wiley, 1992.

Yepsen, Roger B. *Trees for the Yard, Orchard, and Woodlot.* Emmaus, PA: Rodale, 1976.

WINDBREAKS AND BUFFERS

Agriculture Canada. *Best Management Practices: Farm Forestry and Habitat Management.* Toronto: Ontario Federation of Agriculture, 1994.

———. *Best Management Practices: Fish and Wildlife Habitat Management.* Toronto: Ontario Federation of Agriculture, 1996.

Landowner Resource Centre. "The Benefits of Windbreaks" (fact sheet). Manotick, ON: Landowner Resource Centre, 1994.

———. "Designing and Caring for Windbreaks" (fact sheet). Manotick, ON: Landowner Resource Centre, 1994.

CHRISTMAS TREES

Benyus, Janine M. *The Christmas Tree Pest Manual.* Washington: U.S. Department of Agriculture, 1983.

Hill, Lewis. *Christmas Trees: Growing and Selling Trees, Wreaths and Greens.* Pownall, VT: Storey Communications, 1989.

McEvoy, Thomas J. *Using Fertilizers in the Culture of Christmas Trees.* Hinesburg, VT: Paragon Books, 1992.

MAPLE SYRUP

Coons, C.F. *Sugar Bush Management for Maple Syrup Producers.* Toronto: Ministry of Natural Resources, 1983.

Lawrence, James, and Rux Martin. *Sweet Maple.* Shelburne, VT: Chapters Publishing, 1993.

Mann, Rick. *Backyard Sugarin'.* Woodstock, VT: Countryman Press, 1991.

TRAIL BUILDING

Ashbaugh, B.L., and R.F. Holmes. *Trail Planning and Layout.* New York: National Audubon Society, 1967.

Demrow, C., and D. Salisbury. *The Complete Guide to Trail Building and Maintenance.* Boston: Appalachian Mountain Club Books, 1998.

Forest Service. *Trails Management Handbook.* Washington: U.S. Department of Agriculture, 1985.

PESTS AND DISEASES

Armstrong, J.A., and W.G.H. Ives. *Forest Insect Pests in Canada.* Ottawa: Natural Resources Canada, 1995.

Boyce, J.S. *Forest Pathology*. New York: McGraw Hill, 1961.

Hepting, G.H. *Diseases of Forest and Shade Trees of the United States* (handbook #368). Washington: U.S. Department of Agriculture, 1971.

McGauley, B.H., and C.S. Kirby. *Common Pests of Trees in Ontario*. Toronto: Ministry of Natural Resources, 1991.

Myren, D.T. *Three Diseases of Eastern Canada*. Ottawa: Natural Resources Canada, 1994.

Tattar, T.A. *Diseases of Shade Trees*. New York: Academic Press, 1978.

Buying Rural Land

Boudreau, E. *Buying Country Land*. New York: Collier, 1973.

Vardaman, J.M. *How to Make Money Growing Trees*. New York: Wiley, 1989.

Long-Term Conservation

Haney, H.L., and W.C. Siegel. *Estate Planning for Forest Landowners*. New Orleans: Forest Service, Southern Forest Experimental Station, 1993.

Hoose, Phil. *Building an Ark: Tools for the Preservation of Natural Diversity*. Covello, CA: Island Press, 1981.

Landowner Resource Centre. *Making Your Woodland Pay: Financial Aspects of Property Management*. Manotick, ON: Landowner Resource Centre, 1997.

Milne, J. *The Landowner's Options: A Guide to the Voluntary Protection of Land in Maine*. Bangor, ME: Maine Critical Areas Program, 1985.

Mogan, P.J., and G. Nemeth. *Family Trusts for Tax and Estate Planning*. North Vancouver, BC: Self-Counsel Press, 1996.

Nature Writing

Beston, Henry. *The Outermost House*. New York: Ballantine, 1928.

———. *Northern Farm*. New York: Ballantine, 1928.

Brooks, Paul. *Speaking for Nature*. San Francisco: Sierra Club, 1980.

Dillard, Annie. *Pilgrim at Tinker Creek*. New York: Harper and Row, 1974.

De Kiriline Lawrence, Louise. *To Whom the Wilderness Speaks*. Toronto: McGraw-Hill Ryerson, 1980.

Leopold, Aldo. *A Sand County Almanac*. New York: Ballantine, 1949.

Mitchell, John H. *A Field Guide to Your Own Back Yard*. New York: Norton, 1985.

FIELD GUIDES

Farrar, J.L. *Trees in Canada.* Toronto: Fitzhenry and Whiteside, 1995.

Hilts, S. *A Pocket Guide to Ontario Trees and Some Woodland Plants.* Guelph, ON: University of Guelph, 1997.

Leopold, D.J., W.C. McComb, and R.N. Muller. *Trees of the Central Hardwood Forests of North America.* Portland, OR: Timber Press, 1998.

Petrides, G.A. *A Field Guide to Trees and Shrubs.* Boston: Houghton Mifflin, 1986.

Soper, J.H., and M.L. Heimburger. *Shrubs of Ontario.* Toronto: Royal Ontario Museum, 1982.

Symonds, George W.D. *The Shrub Identification Book.* New York: Quill, 1963.
———. *The Tree Identification Book.* New York: Quill, 1958.

A bibliography of further technical sources can be obtained by writing to the authors at the first address below.

Contact Addresses for Publications

Centre for Land and Water Stewardship
University of Guelph
Guelph, ON N1G 2W1
Web page: http://www.uoguelph.ca/CLAWS

Cornell Cooperative Extension Service
Resource Center-MW
7 Business and Technology Park
Ithaca, NY 14850
Web page: http://www.cce.cornell.edu/publications

Landowner Resource Centre
P.O. Box 599
Manotick, ON K4M 1A5
Web page: http://www3.sympatico.ca/lrc

Minnesota's Bookstore
117 University Avenue, J
St. Paul, MN 55155

Pennsylvania State University
College of Agriculture
University Park, PA 16802-2801

University of Vermont Extension Service
University of Vermont
Burlington, VT 05405

Contact Addresses for Professional Assistance

NAME AND ADDRESS	PHONE, FAX & INTERNET	EMPHASIS	OWNERS' ASSOCIATION
CONNECTICUT			
Mr. Stephen H. Broderick *Extension Forester* *Extension Office* *139 Wolf Den Road* *Brooklyn, CT 06234*	*Phone: (860) 744-9600* *Fax: (860) 744-9480* *sbroderi@canr1.cag.* *uconn.edu*	*Forest management, forest-wildlife management, volunteer training*	*Connecticut Forest and Park Association* *16 Meriden Road* *Rockfall, CT 06481* *(860) 346-2372*
Mr. Robert M. Ricard *Cooperative Extension System* *1800 Asylum Avenue* *West Hartford, CT 06117*	*Phone: (860) 570-9257* *Fax: (860) 570-9008* *rricard@canr1.cag.* *uconn.edu*	*Urban forestry, volunteer training*	
ILLINOIS			
Mr. Michael F. Bolin *Extension Forester* *University of Illinois* *W-503 Turner Hall* *1102 South Goodwin Avenue* *Urbana, IL 61801*	*Phone: (217) 333-2778* *Fax: (217) 244-3219* *bolinm@idea.ag.uiuc.edu*	*Forest management, logger education, agroforestry, windbreaks*	*Illinois Woodland Owners and Users Association* *R.R. 1, Box 57* *Mason, IL 62443* *(618) 245-6392*

NAME AND ADDRESS	PHONE, FAX & INTERNET	EMPHASIS	OWNERS' ASSOCIATION

Dr. Gary Rolfe
Department Head
  Forestry
University of Illinois
W-503 Turner Hall
1102 South Goodwin
  Avenue
Urbana, IL 61801
	Phone: (217) 333-2770
	Fax: (217) 244-3219
	g-rolfe@uiuc.edu
		Forest
		management	

INDIANA			

Dr. William L. Hoover
Purdue University
Department of Forestry
  and Natural Resources
West Lafayette, IN
47907-1159
	Phone: (765) 494-3580
	Fax: (765) 496-2422
	billh@forest1.fnr.
	  purdue.edu
		Wood-products
		marketing,
		timber taxation,
		forest economics
			Indiana Forestry and
			  Woodland Owners
			  Association
			2505 Radcliffe Avenue
			Indianapolis, IN 46227
			(317) 888-8484

Mr. Ronald L. Rathfon Southern Indiana Purdue Agricultural Center 11371 Purdue Farm Road DuBois, IN 47527	Phone: (812) 678-3401 Fax: (812) 678-3412	Forest management, hardwood silviculture, land-use planning	
Mr. John R. Seifet Southeast Purdue Agricultural Center Box 155 Butlerville, IN 47223	Phone: (812) 458-6978 Fax: (812) 458-6979	Forest management, hardwood silviculture, reforestation	
MAINE			

Mr. William D. Lilley
Forestry Specialist
University of Maine
Cooperative Extension
  Service
5755 Nutting Hall
Orono, ME 04469-5755
	Phone: (207) 581-2891
	Fax: (207) 581-2890
	blilley@umce.umext.
	  maine.edu
		Forest
		management
			Small Woodland
			  Owners Association
			  of Maine
			P.O. Box 926
			Augusta, ME
			04332-0926
			(207) 626-0005

NAME AND ADDRESS	PHONE, FAX & INTERNET	EMPHASIS	OWNERS' ASSOCIATION

Mr. James F. Philp
Forestry Specialist
University of Maine
Cooperative Extension
  Service
5755 Nutting Hall
Orono, ME 04469-5755
	Phone: (207) 581-2885
	Fax: (207) 581-3466
	jphilp@umce.umext.mai
	  ne.edu		

MASSACHUSETTS			

Mr. Scott Jackson
University of
  Massachusetts
Department of Forestry
  and Wildlife
  Management
301A Holdworth Hall
Amherst, MA 01003
	Phone: (413) 545-4743
	Fax: (413) 545-4358
	sjackson@umext.umass.
	  edu
		Environment
		conservation,
		wetlands,
		wildlife
			Massachusetts
			  Forestry Association
			P.O. Box 1096
			Belchertown, MA
			01007-1096
			(413) 323-7326

Dr. David B. Kittredge
University of
  Massachusetts
Department of Forestry
  and Wildlife
  Management
Holdworth Hall
Amherst, MA 01003
	Phone: (413) 545-2943
	Fax: (413) 545-4358
	dbk@forwild.umass.edu
		Forest
		stewardship,
		forest
		management,
		forest ecology	

MICHIGAN			

Dr. Melvin R. Koelling
Michigan State University
Department of Forestry
126 Natural Resources
  Building
East Lansing, MI
48824-1222
	Phone: (517) 355-0096
	Fax: (517) 432-1143
	Koelling@pilot.msu.edu
		Forest
		management,
		Christmas trees
			Michigan Forestry
			  Association
			1558 Barrington Street
			Ann Arbor, MI
			48103
			(313) 655-8279

Mr. Dean R. Solomon *Michigan State University* *Kellogg Biological Station* *3700 East Gull Lake Drive* *Hickory Corners, MI* *49060*	*Phone: (616) 671-2412* *Fax: (616) 671-2165*	*Forest* *management*	

NAME AND ADDRESS	PHONE, FAX & INTERNET	EMPHASIS	OWNERS' ASSOCIATION
MINNESOTA			
Extension Forest Resources Department of Forest Resources University of Minnesota 1530 Cleveland Avenue North St. Paul, MN 55108	Phone: (612) 624-7222 Fax: (612) 625-5212		Minnesota Forestry Association P.O. Box 496 Grand Rapids, MN 55744 (218) 326-3000
NEW HAMPSHIRE			
Ms. Karen Bennett Extension Specialist Forest Resources University of New Hampshire Cooperative Extension Service 55 College Road Pettee Hall Durham, NH 03824-3599	Phone: (603) 862-4861 Fax: (603) 862-0107 karen.bennett@unh.edu	Forest stewardship, community forestry, environmental education	New Hampshire Timberland Owners Association 54 Portsmouth Street Concord, NH 03301 (603) 224-9699
NEW YORK			
Dr. Peter Smallidge Cornell University Department of Natural Resources 116 Fernow Hall Ithaca, NY 14853-3001	Phone: (607) 255-4696 Fax: (607) 255-2815 pjs23@cornell.edu	Forest management, forest stewardship, forest ecology	New York Forest Owners Association P.O. Box 180 Fairport, NY 14450 (716) 377-6060
OHIO			
Dr. Randy Heiligmann State Forestry Extension Specialist Ohio State University 2021 Coffey Road Columbus, OH 43210	Phone: (614) 292-9838 Fax: (614) 292-7432 heiligmann.1@osu.edu	Forest management, Christmas-tree production, maple-syrup production	Ohio Forestry Association 1335 Dublin Road Suite 203D, Columbus, OH 43215 (614) 486-6767

NAME AND ADDRESS	PHONE, FAX & INTERNET	EMPHASIS	OWNERS' ASSOCIATION
PENNSYLVANIA			
Dr. James C. Finley Penn State University School of Forest Resources 2B Ferguson Building University Park, PA 16802	Phone: (814) 863-0401 Fax: (814) 865-3725 fj4@psu.edu	Forest management, forest stewardship	Pennsylvania Forestry Association 56 East Main Street Mechanicsburg, PA 17055 (717) 766-5371
VERMONT			
Mr. Thom J. McEvoy Extension Forester University of Vermont 340 Aiken Center School of Natural Resources Burlington, VT 05405	Phone: (802) 656-2913 Fax: (802) 656-8683 tmcevoy@together.net	Forest management, logger education, forest ecology	Vermont Woodlands Association P.O. Box 254 Peacham, VT 05862 (802) 584-3333
WISCONSIN			
Dr. A. Jeff Martin University of Wisconsin Department of Forest Ecology and Management 1630 Linden Drive Madison, WI 53706	Phone: (608) 262-0134 Fax: (608) 262-9922 ajmartin@facstaff.wisc. edu	Forest management	Wisconsin Woodland Owners Association P.O. Box 285 Stevens Point, WI 54481 (715) 346-4798

ONTARIO

GOVERNMENT	NONGOVERNMENT
Ontario Ministry of Natural Resources Natural Resources Information Centre P.O. Box 7000 300 Water Street Peterborough, ON K9J 8M5 (416) 314-2000	Ontario Forestry Association 200 Consumers Road Suite 307 North York, ON M2J 4R4 Phone (416) 493-4565
	Ontario Woodlot Association 275 County Road 44 R.R. 4 Kemptville, ON K0G 1J0 Phone: (613) 258-0110

NEW BRUNSWICK

GOVERNMENT	NONGOVERNMENT
Forest Extension Service	*New Brunswick Federation of Woodlot Owners*
Natural Resources and Energy	*180 St. John Street*
P.O. Box 6000	*Fredericton, NB*
Fredericton, NB	*E3B 4A9*
E3B 5H1	*Phone: (506) 459-2990*
Phone: (506) 453-3711	

QUEBEC

GOVERNMENT	NONGOVERNMENT
Forêt Québec	*Fédération des producteurs de bois du Québec*
Service de la Mise en valeur de la forêt privée	*555, boulevard Roland-Therrien*
880, chemin Ste-Foy, 5e étage	*Longueuil (Québec)*
Québec (Québec)	*J4H 3Y9*
G1S 4X4	*Phone: (450) 679-0530*
Phone: (418) 644-2194	*Fax: (450) 679-5682*
Fax: (418) 646-9245	
	Le Regroupement des sociétés d'aménagement
	forestier du Québec
	3405-C, boulevard Wilfrid-Hamel, bureau 330
	Québec (Québec)
	G1P 2J3
	Phone: (418) 855-1344
	Fax: (418) 877-6449

Addresses are correct to the best of our knowledge. Most are taken from *1996 Conservation Directory* of the National Wildlife Federation (U.S.) or from *Cooperative Extension Service Personnel in Forest Management and Wood Products,* July 1997–98, published by the USDA.

INDEX